Digital Ethics

This fifth volume in Christian Fuchs's *Media, Communication and Society* series presents foundations and applications of digital ethics based on critical theory. It applies a critical approach to ethics within the realm of digital technology.

Based on the notions of alienation, communication (in)justice, media (in)justice, and digital (in)justice, it analyses ethics in the context of digital labour and the surveillance-industrial complex; social media research ethics; privacy on Facebook; participation, co-operation, and sustainability in the information society; the digital commons; the digital public sphere; and digital democracy. The book consists of three parts. Part I presents some of the philosophical foundations of critical, humanist digital ethics. Part II applies these foundations to concrete digital ethics case studies. Part III presents broad conclusions about how to advance the digital commons, the digital public sphere, and digital democracy, which is the ultimate goal of digital ethics.

This book is essential reading for both students and researchers in media, culture, communication studies, and related disciplines.

Christian Fuchs is Chair Professor of Media Systems and Media Organisation at Paderborn University, Germany. His fields of expertise are critical digital and social media studies, Internet and society, the political economy of media and communication, information society theory, social theory, and critical theory. He is the author of numerous publications in these fields.

Digital Ethics
Media, Communication and Society
Volume Five

Christian Fuchs

Routledge
Taylor & Francis Group

LONDON AND NEW YORK

Cover image: mammuth, Getty Images

First published 2023
by Routledge
4 Park Square, Milton Park, Abingdon, Oxon OX14 4RN

and by Routledge
605 Third Avenue, New York, NY 10158

Routledge is an imprint of the Taylor & Francis Group, an informa business

British Library Cataloguing-in-Publication Data
A catalogue record for this book is available from the British Library

Library of Congress Cataloging-in-Publication Data
Names: Fuchs, Christian, 1976- author.
Title: Digital ethics / Christian Fuchs.
Description: Abingdon, Oxon; New York, NY: Routledge, 2023. |
Series: Media, communication and society; volume five |
Includes bibliographical references and index.
Identifiers: LCCN 2022010363 (print) | LCCN 2022010364 (ebook) |
ISBN 9781032246147 (hardback) | ISBN 9781032246161 (paperback) |
ISBN 9781003279488 (ebook)
Subjects: LCSH: Internet—Moral and ethical aspects. |
Internet—Social aspects. | Information society.
Classification: LCC TK5105.878 .F83 2023 (print) |
LCC TK5105.878 (ebook) | DDC 395.5—dc23/eng/20220625
LC record available at https://lccn.loc.gov/2022010363
LC ebook record available at https://lccn.loc.gov/2022010364

ISBN: 978-1-032-24614-7 (hbk)
ISBN: 978-1-032-24616-1 (pbk)
ISBN: 978-1-003-27948-8 (ebk)

DOI: 10.4324/9781003279488

Typeset in Univers
by codeMantra

Contents

Figures

Tables

Acknowledgements

Chapter 2 was first published as a journal article using a Creative Commons CC-BY licence that allows reprint: Christian Fuchs. 2021. Foundations of Communication/Media/Digital (In)Justice. *Journal of Media Ethics* 36 (4): 186–201. http://doi.org/10.1080/2373 6992.2021.1964968

A shorter version of Chapter 3 was first published as a journal article. The chapter is an extended version of the original paper. It has been reused and extended based on a contractual stipulation in the author agreement with Taylor & Francis that allows republication and modification. Christian Fuchs. 2020. The Ethics of the Digital Commons. *Journal of Media Ethics* 35 (2): 112–126. https://doi.org/10.1080/23736992.2020.1736077

Chapter 4 was first published as a book chapter. It has been reprinted based on a stipulation in the author agreement that enables the author to reprint his chapter in a volume of his own works. Fuchs, Christian. 2016. Information Ethics in the Age of Digital Labour and the Surveillance-Industrial Complex. In *Information Cultures in the Digital Age: A Festschrift in Honor of Rafael Capurro*, ed. Matthew Kelly and Jared Bielby, 173–190. Wiesbaden: Springer.

Chapter 5 was first published as a book chapter. It has been reprinted based on a stipulation in the author agreement that enables the author to reprint his chapter in a volume of his own works. Fuchs, Christian. 2018. "Dear Mr. Neo-Nazi, can you please give me your informed consent so that I can quote your fascist tweet?": Questions of Social Media Research Ethics in Online Ideology Critique. In *The Routledge Companion to Media and Activism*, ed. Graham Meikle, 385–394. Abingdon: Routledge.

Chapter 6 was first published as a journal article. It has been reprinted based on Emerald's Author Rights (https://www.emeraldgrouppublishing.com/our-services/authors/author-policies/author-rights) Fuchs, Christian. 2011. Towards an Alternative Concept of Privacy. *Journal of Information, Communication and Ethics in Society* 9 (4): 220–237. https://doi.org/10.1108/14779961111191039

Chapter 7 was first published as a journal article. It has been reprinted based on SAGE's Author Archiving and Re-Use Guidelines (https://uk.sagepub.com/en-gb/eur/journal-author-archiving-policies-and-re-use): Fuchs, Christian. 2012. The Political Economy of Privacy on Facebook. *Television & New Media* 13 (2): 139–159. https://doi.org/10.1177%2F1527476411415699

Chapter 8 was first published as a journal article. It has been reprinted with permission of *International Journal of Communication*. Fuchs, Christian. 2017. Information Technology and Sustainability in the Information Society. *International Journal of Communication* 11: 2431–2461. Published open access: https://ijoc.org/index.php/ijoc/article/view/6827

Chapter 9 was first published as a journal article. It has been reprinted based on a contractual stipulation in the author agreement with Taylor & Francis that allows republication. Christian Fuchs. 2010. Theoretical Foundations of Defining the Participatory, Co-operative, Sustainable Information Society (PCSIS). *Information, Communication, and Society* 13 (1): 23–47. https://doi.org/10.1080/13691180902801585

Chapter 10 was first published as a journal article using a Creative Commons CC-BY licence that allows reprinting it. Christian Fuchs. 2021. The Digital Commons and the Digital Public Sphere: How to Advance Digital Democracy Today. *Westminster Papers in Communication and Culture* 16 (1): 9–26. https://doi.org/10.16997/wpcc.917

Part I

Foundations

Chapter One
What Is Digital Ethics?

1.1 About this Book

This book asks: what does digital ethics look like when it is based on the critical theory of Marxist humanism? What are the principles of critical, Marxist-humanist ethics and how can this approach be applied for explaining digital society's moral principles and practices?

The book at hand is the fifth volume of a series of books titled "*Media, Communication and Society*". The overall aim of *Media, Communication & Society* is to outline the foundations of a critical theory of communication and digital communication in society. It is a multi-volume book series situated on the intersection of communication theory, sociology, and philosophy. The overall questions that "Media, Communication & Society" deals with are: what is the role of communication in society? What is the role of communication in capitalism? What is the role of communication in digital capitalism?

Based on critical theory and Marxist humanism, this book presents foundations and applications of digital ethics. It combines the approaches of Aristotle, Karl Marx, and Alasdair MacIntyre and applies this combination to the realm of digital technology. The book outlines based on Marx's notion of alienation principles communication (in)justice, media (in)justice, and digital (in)justice. It analyses the digital commons, ethics in the context of digital labour and the surveillance-industrial complex, social media research ethics, socialist privacy, privacy on Facebook, participation, co-operation and sustainability in the information society, the digital commons, the digital public sphere, and digital democracy.

The book consists of three parts. Part I (Chapters 1–3) presents philosophical foundations of critical, Marxist-humanist digital ethics. Part II applies these foundations to concrete digital ethics case studies (Chapters 4–9). Part III (Chapter 10) presents broad

DOI: 10.4324/9781003279488-2

conclusions about how to advance the digital commons, the digital public sphere, and digital democracy, which is the penultimate goal of digital ethics.

The three key authors whom you will encounter in this book are Aristotle, Karl Marx, and Alastair MacIntyre.

Aristotle (384-322 BC) was a Greek philosopher. He founded the Aristotelian tradition of philosophy, of which virtue ethics forms one part. Among Aristotle's most well-known books are *Metaphysics*, *Physics*, *De Anima*, *Politics*, *Nicomachean Ethics*, and *Eudemian Ethics*. Aristotle influenced philosophers such as Averroes, Avicenna, Thomas Aquinas, Francis Bacon, Nicolaus Copernicus, René Descartes, Georg Wilhelm Friedrich Hegel, Karl Marx, Hannah Arendt, Alasdair MacIntyre, Martha Nussbaum, and Michael Sandel.

Karl Marx (1818–1883) was a philosopher, economist, sociologist, journalist, and revolutionary socialist. In 1999, he won a BBC online poll that determined the millennium's "greatest thinker" (BBC 1999). His key works include *Economic and Philosophic Manuscripts*, *The Manifesto of the Communist Party* (together with Friedrich Engels), *Grundrisse*, and the three volumes of *Capital*. Karl Marx plays a role throughout this book in all chapters.

Alasdair C. MacIntyre (born in 1929) is a philosopher and the most well-known and most influential contemporary Aristotelian philosophers. His thought was especially influenced by Aristotle, Karl Marx, and Thomas Aquinas. MacIntyre argues that Aristotelian ethics has been rather forgotten and ignored. His task is to renew Aristotelian moral philosophy. Among MacIntyre's books are *Marxism: An Interpretation*, *A Short History of Ethics*, *Marxism and Christianity*, *After Virtue*, *Dependent Rational Animals: Why Human Beings Need the Virtues*, and *Ethics in the Conflicts of Modernity: An Essay on Desire, Practical Reasoning, and Narrative*.

Each chapter focuses on illuminating answers to a specific question:

Chapter 1: What is ethics? What is digital ethics?

Chapter 2: What are communication justice, media justice, and digital justice and how can they be studied based on Marxist humanism?

Chapter 3: Why is it morally good to foster the digital commons and how can we ethically justify the importance of the digital commons?

Chapter 4: How can Rafael Capurro's information ethics inform the critical analysis of digital labour and the surveillance-industrial complex?

Chapter 5: What kind of research ethics do we need in Internet and social media research?

Chapter 6: What is a critical concept of privacy?

Chapter 7: What ethical aspects are there of Facebook?

Chapter 8: How can we think of sustainability and ICTs in the context of a critical theory of society? How is the sustainability of ICTs related to capitalism and class?

Chapter 9: What is a participatory, co-operative, sustainable information society?

Chapter 10: What are the democratic potentials of the digital commons and the digital public sphere?

1.2 What Is Ethics?

If we want to understand what digital ethics is about, we need to understand what ethics is. Ethics studies the principles of morality and moral action. Morality refers to that which is considered right and wrong. The Internet *Encyclopedia of Philosophy* defines ethics in the following manner:

"The field of ethics (or moral philosophy) involves systematizing, defending, and recommending concepts of right and wrong behavior. Philosophers today usually divide ethical theories into three general subject areas: metaethics, normative ethics, and applied ethics. *Metaethics* investigates where our ethical principles come from, and what they mean. Are they merely social inventions? Do they involve more than expressions of our individual emotions? Metaethical answers to these questions focus on the issues of universal truths, the will of God, the role of reason in ethical judgments, and the meaning of ethical terms themselves. *Normative ethics* takes on a more practical task, which is to arrive at moral standards that regulate right and wrong conduct. This may involve articulating the good habits that we should acquire, the duties that we should follow, or the consequences of our behavior on others. Finally, *applied ethics* involves examining specific controversial issues, such as abortion, infanticide, animal rights, environmental concerns, homosexuality, capital punishment, or nuclear war".

(Fieser 2005)

Ethics is moral philosophy focused on the study of the moral foundations of society, moral principles, and moral practices. Digital ethics is an ethics that studies the foundations of digital society and the principles and practices of morality in the context of digitalisation. It is concerned with principles and practices of how humans should act in light of the problems and challenges that digitalisation poses for society. Morality has to do with what one "should do in that or relevantly similar situations" (MacIntyre 1957, 331), i.e. what one ought to do. Moral judgements employ "ought" in their language (MacIntyre 1957, 331)

Applied ethics involve, for example, health care and medical ethics, bioethics, social ethics, and political ethics; medical ethics, bioethics, environmental ethics, and business ethics are examples of applied ethics (Dittmer 2013). Is digital ethics also a version of applied ethics? On the one hand, the answer is "yes" because digital ethics studies morality in the context of computing and digital technologies. On the other hand, the answer is "no" because almost all aspects of society have been digitalised and are shaped by digital technologies, which is why moral questions of digitalisation are of relevance for society as a whole and are thereby part of general ethics. Digital ethics is both a form of applied and general ethics that impacts the ethics of society and requires specialised insights for professionals who engineer, produce, and work with digital technologies and digital content.

1.2.1 Four Forms of Ethics: Virtue Ethics, Deontology, Consequentialism, Critical Ethics

Virtue ethics, deontology, and consequentialism are three major approaches in ethics (Fieser 2005). One can attempt to justify the moral superiority of digital commons over digital commodities based on any of these and other approaches (see Table 1.1). In typologies of ethical approaches, critical ethics that is based on the Marxian approach is mostly missing. Given that the book *Digital Ethics* wants to contribute to the advancement of a Marxist-humanist ethics, we are adding critical ethics understood as Marxist-humanist ethics to a typology of important ethics approaches (Table 1.1).

1.2.2 Virtue Ethics: Aristotle

Virtue ethics goes back to Aristotle. It deals with how to achieve and design education that brings about moral beings that act in manners that bring about happiness and a good society. For Aristotle (2002, §1007a), a virtue is "an active condition that makes one apt at choosing" between options so that society can achieve "what is best and what is done well". Virtues pertain "either to thinking or to character" (§1103 a) and have to do with "living well and acting well" (§1098b) so that happiness is advanced. Aristotle (2013b, §§1220b-1221a) identifies 14 virtues: mildness, courage, shame, temperance, righteous indignation, justice, liberality, truth, friendliness, dignity, endurance, magnanimity, magnificence, and prudence.

TABLE 1.1 The application of four ethical approaches to digitalisation

Approach	Description	Application to the digital
Virtue ethics	This approach says that humans should strive for and practice moral excellence in order to create a good individual and community life. It therefore argues for moral education in the development of virtues such as wisdom, courage, temperance, justice, fortitude, generosity, self-respect, good temper, or sincerity.	Fostering digital commons and a good digital society is the realisation of the virtue of creating the common good in the digital sphere and in digital society. Humans should develop virtues in respect to digitalisation.
Deontology	Humans have individual rights and duties that should guide their behaviour. Kant's Golden Rule: Treat others like you want to be treated by others.	If one expects to be treated well by others in the online world and digital society, then one should treat others well online and in the digital society.
Consequentialism	An action is morally right or morally preferable over another action if its consequences produce more good than harm in comparison to other actions.	If a certain form of digitalisation as a net total produces more advantages than disadvantages, then this form of digitalisation should be implemented and advanced.
Critical theory as Marxist-humanist ethics	A society is a true society if it advances the good life for all and a false society if is a class society where some benefit at the expense of others. Humans should struggle against conditions that are opposed to the good life for all, i.e. class relations and domination.	There should be struggles against and opposition to digitalisation that advances exploitation and domination and support for and advancement of digitalisation that fosters the common good that benefits all.

1.2.3 Deontological Ethics: Kant

Deontology has to do with duties. The term "deontology" comes from the Greek word δέον that means duty. It is a kind of ethics that does not focus on the consequences of action, but on the question of whether those acting have the right kind of motives. It wants to identify principles that guide moral action. Immanuel Kant is the most well-known representative of deontological ethics. He based the development of his moral philosophy on the German philosopher Samuel Pufendorf. Kant's ethics is based on an absolute rule that is termed the Golden Rule. It says: treat others like you want to be treated by them.

> "Act only according to that maxim by which you can at the same time will that it should become a universal law. […] Act as though the maxim of your action were by your will to become a universal law of nature. […] So act that you use humanity, whether in your own person or in the person of any another, always at the same time as an end, never merely as a means".
>
> (Kant 2011, 71, 87)

The Golden Rule is an absolute ethical principle. It understands itself as a rule of conduct applicable to any situation. For Kant, the Golden Rule is the embodiment of freedom and a principle for advancing freedom. Jürgen Habermas (2008, 140) argues that Kant's categorical imperative is reflected in the insight that freedoms are only limited by the freedom of others. Habermas (2011, 14) says that Kant's principle of autonomy and his categorical imperative is present in the *Universal Declaration of Human Rights'* §1: "All human beings are born free and equal in dignity and rights".

The Golden Rule fails in situations where people are willing to suffer, tolerate violence against themselves, or die if they were in the positions of others. Kant's ethics is transcendental in the sense that it is grounded in the category of freedom as the highest and absolute principle. For Kant, moral freedom means that humans resist their instincts and desires and hence restrict absolute freedom of action by giving themselves rules of conduct that enable true freedom. The Categorical Imperative is considered as an expression of freedom, good will would be oriented on freedom. Another absolute rule is the Rule of Golden Mean by Aristotle which says that happiness can be found by choosing the middle way between extremes.

1.2.4 Consequentialism: Bentham

Consequentialism is a form of ethics, in which the moral rightness or wrongness of an action is determined by its consequences. An action is morally right if, on the whole, its consequences are more positive than negative. Jeremy Bentham's utilitarian ethics is the best-known version of consequentialist ethics.

For Bentham (2000), utility has to do with pain and pleasure (14). Action can "augment or diminish the happiness of the party whose interest is in question" (14). Bentham considers an action morally good if it increases the happiness for the largest number of people in a community.

> "The general tendency of an act is more or less pernicious, according to the sum total of its consequences: that is, according to the difference between the sum of such as are good, and the sum of such as are evil".
>
> (61)

He gives a utilitarian definition of ethics: "Ethics at large may be defined, the art of directing men's actions to the production of the greatest possible quantity of happiness, on the part of those whose interest is in view" (225).

"Now private ethics has happiness for its end: and legislation can have no other. Private ethics concerns every member, that is, the happiness and the actions of every member, of any community that can be proposed; and legislation can concern no more".

(227)

A common criticism of consequentialism is that it violates equality and personal rights and can advance inhumanity: Consequentialism

"is not egalitarian because it does not care whether happiness is distributed equally or unequally among people. If the greatest total can be created only by exploiting the miserable to make the happy even happier, then such consequentialism would seem to say that you should do it. [...] Consequentialism may ask us to meddle too much into other people's business. [...] suppose that by using your grandmother's pension to contribute to efficient and thoughtful charities you can develop permanent clean water supplies for many distant villages, thus saving hundreds of people from painful early deaths and permitting economic development to begin. You need only keep her bound and gagged in the cellar and force her to sign the checks. Consequentialism would seem to say that you should do this, but moral common sense says that you should not. [...] Consequentialism seems to tell us to make all our decisions by thinking about overall consequences. But that way of thinking about life is, one might think, inhuman and immoral. When someone asks you a question, you should not stop to calculate the consequences before deciding whether to answer truthfully. If you decide by looking to the consequences, you are not really an honest person".

(Haines 2006)

1.2.5 Critical Theory as Marxist-Humanist Ethics

Marx and Engels see ethics and morality as an expression of a society that "has hitherto moved in class antagonisms" (Engels 1878, 87), which is why for them "morality has always been class morality" (87). But they also speak of a humane morality that transcends class society: "A really human morality which stands above class antagonisms and above any recollection of them becomes possible only at a stage of society which has not only overcome class antagonisms but has even forgotten them in practical life" (88).

Orthodox interpretations of Marx and Engels argue that their approach is science that studies the laws of capitalism. They often expect that capitalism automatically collapses and underestimate the importance of collective praxis in social change. Societal conditions shape but do not determine collective human action. If an individual and groups act in a certain manner in a situation, such as a crisis of society, is not determined but depends on how their social relations shape their worldviews, including their moral judgements. Marx's own theory is full of moral language and moral judgement that condemn exploitation and class society. He formulates an ethical imperative, namely, that humans should overthrow exploitation and domination through class and social struggles: critical theory's "*categorical imperative*" to

> "*overthrow all relations* in which man is a debased, enslaved, forsaken, despicable being, relations which cannot be better described than by the exclamation of a Frenchman when it was planned to introduce a tax on dogs: Poor dogs! They want to treat you like human beings!"
>
> (Marx 1844, 182)

Critical theory is a critical ethics that analyses class and power relations, ideology, and social struggles with the goal in mind to inform struggles for a society that enables and implements the good life for all humans, a humanist and socialist society.

Marxian ethics can in principle be combined with a variety of other ethical approaches. Aristotelian thought has influenced Marx in a variety of respects (see McCarthy 1992; Meikle 1985; Pike 2018), including ethics. On the one hand, critical thought and practices that question and struggle against exploitation and domination and for socialism constitute virtuous praxis. Critical pedagogy is therefore an Aristotelian-Marxian form of critical education. On the other hand, Marx shares and builds on Aristotle's notion of the common good. Socialism is a society that benefits all, which requires to establish a classless society. Marx's theory "involved a reworking of Aristotle's ethics through Hegelian lenses" and "this synthesis was made from the standpoint of workers' struggles against capital" (Blackledge 2012, 15). He developed a "critique of existing social relations from the point of view of the struggle for human freedom" (Blackledge 2012, 15)

Aristotle analyses the commons in the context of the virtues of friendship and justice:

> "To whatever extent that they share something in common, to that extent is there a friendship, since that too is the extent to which there is something just. And the proverb 'the things of friends are common' is right, since friendship consists in community".
>
> (Aristotle 2002, §1159b)

The "common advantage" is called "just" (1160a). Democracies in contrast to tyrannies advance the common good: "So friendships and justice are of small extent in tyrannies, but in democracies they are of greater extent, since many things are common to people who are equal" (§1161b).

Aristotle (2002) discusses the violation of the common good in the context of the virtue of generosity. It "is not characteristic of someone who does good for other to have a ready hand for taking benefits from them" (§1120a). Wasteful people are stingy in that "they are driven to provide for themselves from other sources" (§1121a). They "take money carelessly and from everywhere" (§1121b). Aristotle (2002) discusses justice and injustice in book V of the *Nicomachean Ethics*. He stresses that justice can either be understood as that which is lawful or that which is equitable, but that the two are different things. The unjust person is "greedy for more" (§1129b). If "one makes a profit, it is referred to no vice other than injustice" (§1130a). Aristotle distinguishes between distributive, corrective, reciprocal, and universal justice (that advances the common good) (McCarthy 1990, Chapter 2). Justice and injustice are for Aristotle matters of proportionality and disproportionality. An "unjust person has more, while the one to whom injustice is done has less of something good" (Aristotle 2002, §1131b). Injustice means that someone has "an excess for oneself of what is simply beneficial and a deficiency of what is harmful" (1134a). So, Aristotle argues that injustice means that a certain individual, or group has a kind of exclusive control of a good over others.

Aristotle (2002) not just opposes injustice to justice, but also to friendship and love, which are social relations where humans benefit and do good things for others without instrumental interests. The common arises from friendship and community: in

> "every sort of community there seems to be something just, and also friendship. [...] To whatever extent that they share something in common, to that extent is there a friendship, since that too is the extent to which there is something just. And the proverb 'the tings of friends are common' is right, since friendship consist in community. All things are common to brothers and comrades".
>
> (§1159b)

The political community aims at an advantage "that extends to all of life" (§1160a). Aristotle (2013a, §1279a) terms a community where "the multitude governs with a view to the common advantage" polity.

Aristotle (1999, §1048a) sees potentiality as being "capable of something" and being "capable of causing motion". Potency is also the source of dialectic because whatever is potential "is itself capable of opposite effects" (Aristotle 1999, §1051a). In a good society, the full potentials of human beings and society are actualised. In essence, humans are co-operative, social, societal beings, who strive for solidarity and a good life. A particular societal condition enables or hinders the realisation of society's and human potentials. The commons are conditions that aim at the realisation of societal and human potentials.

In an Aristotelian view, the commons are goods that all humans require in order to live a good life. The good life of the individual is only possible in a good society that enables the good life for all. Achieving a good society that benefits all requires the common good. Without being able to live a good life, humans are not fully developed humans and they are denied those common goods that humans and society require to flourish and thereby realise their potentials. In Marx's approach, Aristotle's notion of the common good is reflected in the concept of the human essence, which describes the basic characteristic of humans as social and producing beings, the dialectic of essence and existence that implies critique and the categorical imperatives, and the concept of common property as part of a just society. For example, Marx (1875) speaks of the need for a society wherein "the means of labour are common property and the total labour is collectively regulated" (84) and the high level of productivity combined with common property of the means of production enables practicing the principle "From each according to his abilities, to each according to his needs!" (87).

1.3 What Is Digital Ethics?

Justifying why digital and communication ethics is a task for computer ethics. But what is computer ethics/digital ethics? It is a philosophical field of study where ethics, communication studies, and computer science intersect. Norbert Wiener, Donn Parker, Joseph Weizenbaum, and Walter Maner were some of the early pioneers of computer ethics (Bynum 2001). For Maner (1980), computer ethics studies how computer technology aggravates, transforms, or creates ethical problems.

Rafael Capurro (2005) argues that information and digital ethics deals with ethical questions in the context of the Internet, computer science, the mass media, library and information science, business information, and biological and medical information.

"Information ethics explores and evaluates: the development of moral values in the information field, the creation of new power structures in the information field, information myths, hidden contradictions and intentionalities in information theories and practices, the development of ethical conflicts in the information field".

(Capurro 2005, 7)

James Moor argues that computer ethics deals with the question of

> "how computer technology should be used. Computers provide us with new capabilities and these in turn give us new choices for action. Often, either no policies for conduct in these situations exist or existing policies seem inadequate. A central task of computer ethics is to determine what we should do in such cases, that is, formulate policies to guide our actions".
>
> (Moor 1985, 266)

Computer ethics studies computing's "complex social, ethical, and value concerns" (Johnson 2004, 65). Computer ethics is "the analysis of the nature and social impact of computer technology and the corresponding formulation and justification of policies for the ethical use of such technology" (Moor 1985, 266). Internet ethics (sometimes also termed cyberethics, a term that today sounds somewhat opaque) is about metanorms that guide "acting well in this new realm of cyberspace" (Spinello 2003, 2; see also Ess 2009; Tavani 2011).

In Internet and digital media research, digital ethics and Internet ethics are often reduced to the elaboration and application of principles of research ethics in the conduct of digital methods such as questions of informed consent, privacy, anonymity, etc. Such guidelines are certainly important and an aspect of digital ethics (see Chapter 5 in this book; Townsend et al. 2016), but digital ethics is broader and also covers philosophical and axiological questions of the digital society, which requires an engagement with philosophy from a digital perspective, i.e. the development of a philosophy of the digital and digital society. In *The Handbook of Internet Studies* (Consalvo and Ess 2010), co-edited by digital media ethicist Charles Ess (!), there is one chapter covering Internet research ethics (Chapter 8) but there are no chapters dedicated to philosophical questions of the Internet in digital society. *The Oxford Handbook of Internet Studies* (Dutton 2013) does not contain a chapter about philosophical and ethical aspects of the Internet in society. The terms "philosophy", "ethics", "ontology", and "epistemology" are hardly mentioned on the almost 600 pages of this handbook. In the *International Handbook of Internet Research* (Hunsinger, Klastrup and Allen 2010) and the *Second International Handbook of Internet Research* (Hunsinger, Allen and Klastrump 2020), ethical aspects of the Internet are primarily presented as questions of research ethics. Reduced to research ethics, digital ethics is a positivistic approach devoid of its potential to contribute to the critical analysis of digital society's power structures and inequalities.

A different approach to the ethics and philosophy of the digital, the Internet, and digital society is needed. The Internet and digital technologies are embedded into society's power structures, which brings about moral dilemmas and the question of how we should best assess certain digital technologies' impacts on society. We therefore require a digital ethics that studies the moral aspects of the digital and digital society in the context of power structures.

Digital ethics is an ethics that studies the foundations of digital society and the principles and practices of morality in the context of digitalisation. It is concerned with principles and practices of how humans should act in light of the problems and challenges that digitalisation poses for society. Digital ethics studies principles of what humans ought to do in digital society and in the context of digital technologies and moral practices in digital society and in digital spaces.

Table 1.1 contains a column that shows what the four outlined ethical approaches mean when applied to digitalisation. There is a variety of approaches to digital ethics. Digital virtue ethics stresses how education can strengthen human virtues in respect to digitalisation and that it is virtuous to advance the digital commons. Digital deontological ethics argues that one should treat others well online and in the digital society so that one can expect that others do the same. Digital consequentialist ethics argues that forms of digitalisation should be advanced that as a net total do more good than they cause harm. Critical Marxist-humanist ethics says there should be struggles against and opposition to digitalisation that advances exploitation and domination and support for and advancement of digitalisation that fosters the common good that benefits all. The commonality between Aristotelian digital ethics and critical Marxist-humanist ethics is the focus on the advancement of the digital commons and the common good in digital society.

1.4 The Chapters in this Book

I. Foundations

Chapter 1: What Is Digital Ethics?

> This chapter introduces the book *Digital Ethics*. It provides an overview of the chapters in the book, discusses and defines ethics, and digital ethics. It outlines what a critical theory and Marxist-humanist approach to digital ethics is.

Chapter 2: Foundations of Communication/Media/Digital (In)Justice

> The task of this chapter is to outline the foundations of a Marxist-humanist approach to communication justice, media justice, and digital justice. A dialectical approach to justice is outlined that differs from idealist monism, dualism, and pluralism. It conceives of injustice as alienation and inhumanity and justice as

humanism. This approach is applied to communication, media, and the digital. The chapter outlines concepts and dimensions of (in)justice in general, communication (in)justice, media (in)justice, and digital (in)justice.

Following an introduction, Section 2 engages with theories of justice. Section 3 presents an approach of how to think about alienation as injustice. Section 4 focuses on communication/media (in)justice. Section 5 provides a framework for the analysis of digital (in)justice. Some conclusions are drawn in Section 6.

Chapter 3: The Ethics of the Digital Commons

This chapter asks: why is it morally good to foster the digital commons? How can we ethically justify the importance of the digital commons? An answer is given based on Aristotelian ethics.

Given that the common good plays an important role in Aristotelian ethics, Aristotle's approach is suited for the attempt to ground the ethical foundations of the digital commons. Because Alasdair MacIntyre is the most influential Aristotelian moral philosopher today, the chapter engages with the foundations of MacIntyre's works and gives special attention to his concept of the common good and his analysis of how structures of domination damage the common good.

It is argued that for advancing a philosophy of the (digital) commons, MacIntyre's early and later works, in which he has been influenced by Karl Marx, are of particular importance. The approach taken in this chapter combines Aristotle, Marx, and MacIntyre.

II. Applications

Chapter 4: Information Ethics in the Age of Digital Labour and the Surveillance-Industrial Complex

The rise of computing and the Internet has brought about an ethical field of studies that some term information ethics, computer ethics, digital media ethics, or Internet ethics. The aim of this chapter is to discuss information ethics' foundations in the context of the Internet's political economy. The chapter first looks to ground the analysis in a comparison of two information ethics approaches, namely those outlined by Rafael Capurro and Luciano Floridi. It then develops, based on these foundations, analyses of the information ethical dimensions of two important areas of social media: one concerns the framing of social media by a surveillance-industrial complex in the context of Edward Snowden's revelations and the other deals with issues of digital labour processes and issues of class that arises in this context. The chapter asks ethical questions about these two phenomena that bring up issues of power, exploitation, and control in the information age. It asks if, and if so, how, the approaches of Capurro and Floridi can help us to understand ethico-political aspects of the surveillance-industrial complex and digital labour.

Chapter 5: "Dear Mr. Neo-Nazi, can you please give me your informed consent so that I can quote your fascist tweet?": Questions of Social Media Research Ethics in Online Ideology Critique

Social media is a kind of mirror of what is happening in society. Studying social media content is therefore a good way of studying society. This chapter deals with the question of how to deal with research ethics in qualitative online research. It discusses the limits of established research ethics guidelines. The chapter outlines the foundations of critical-realist Internet research ethics. It provides some examples of how to use a framework. The International Sociological Association's 2001 Code of Ethics argues in respect to informed consent: the security, anonymity, and privacy of research subjects and informants should be respected rigourously, in both quantitative and qualitative research. Debates on Internet research ethics face two extremes. On the one side, research ethics fundamentalism obstructs qualitative online research. On the other, big data positivism lacks a critical focus on qualitative dimensions of analysis.

Chapter 6: Towards an Alternative Concept of Privacy

There are a lot of discussions about privacy in relation to contemporary communication systems (such as Facebook and other "social media" platforms), but discussions about privacy on the Internet in most cases miss a profound understanding of the notion of privacy and where this notion is coming from. The purpose of this chapter is to challenge the liberal notion of privacy and explore the foundations of an alternative privacy conception.

A typology of privacy definitions is elaborated based on Giddens' theory of structuration. The concept of privacy fetishism that is based on critical political economy is introduced. Limits of the liberal concept of privacy are discussed. This discussion is connected to the theories of Marx, Arendt, and Habermas. Some foundations of an alternative privacy concept are outlined.

The notion of privacy fetishism is introduced for criticising naturalistic accounts of privacy. Marx and Engels have advanced four elements of the critique of the liberal privacy concept that were partly taken up by Arendt and Habermas: privacy as atomism that advances; possessive individualism that harms the public good; legitimises and reproduces the capitalist class structure; and capitalist patriarchy.

Given the criticisms advanced in this chapter, the need for an alternative, social-ist privacy concept is ascertained and it is argued that privacy rights should be differentiated according to the position individuals occupy in the power structure, so that surveillance makes transparent wealth and income gaps and company's profits and privacy protects workers and consumers from capitalist domination.

The chapter contributes to the establishment of a concept of privacy that is grounded in critical political economy. Owing to the liberal bias of the privacy concept, the theorisation of privacy has thus far been largely ignored in critical political economy. The chapter contributes to illuminating this blind spot.

Chapter 7: The Ethics and Political Economy of Privacy on Facebook

This chapter provides an analysis of the political economy of privacy and sur-veillance on Facebook. The ideological premises of the liberal privacy concept are criticised and a differentiated concept of consumer privacy is advanced that is applied to the realm of the contemporary Internet, which requires speaking about Internet prosumer privacy. An analysis of the privacy policies of Facebook and other corporate social media platforms shows that targeted advertising is a mechanism that is at the heart of capital accumulation on corporate social media. It is guaranteed by complex legal policies that are based on self-regulatory policy frameworks formulated by Internet companies. Capital accumulation on Facebook is based on the commodification of users and their data. One can in this context speak based on Dallas Smythe of the exploitation of the Internet prosumer com-modity. The convergence of the two realms of play and labour has resulted in the emergence of playbour activities that are exploited by Internet companies in order to maximise profits. Discussions about privacy and surveillance on Facebook and social media in general are situated within the context of the Internet prosumer commodity and online playbour.

Chapter 8: Information Technology and Sustainability in the Information Society

The sustainability concept has developed in a policy context. Its main relevance has been in policy forums such as the United Nations Conference on Environ-ment & Development and the United Nations Conference on Sustainable De-velopment. In the realm of information and communication technologies (ICTs), sustainability has played a policy role in the context of the World Summit on the Information Society (WSIS). This chapter asks: how can we think of sustainability and ICTs in the context of a critical theory of society? How is the sustainability of ICTs related to capitalism and class? It provides a critique of the dominant reductionist and dualistic understandings of information technology sustainability

The Chapters in this Book

in an information society context. The question that arises in this context is if from a critical theory perspective, the sustainability concept should therefore be discarded or not. The view advanced in this chapter is that a critical social theory should provide an ideology critique of information technology sustainability, but at the same time not discard, but transform the sustainability concept into a critical notion of (un)sustainable information technology sustainability in the context of the information society.

Chapter 9: Theoretical Foundations of Defining the Participatory, Co-operative, Sustainable Information Society (PCSIS)

The task of this chapter is to provide a comparative and theoretically grounded discussion of the notions of sustainability, inclusion, and participation in the information society discourse. A theoretical model of society as dialectical system is introduced, in which the economic base and the political–cultural superstructure are mutually shaping each other. Based on a distinction between reductionistic, holistic, dualistic, and dialectical worldviews, four different theoretical approaches on defining the sustainable information society are distinguished, which are based on how the relationship between base and superstructure is conceived. Reductionistic approaches see ecological, technological, or economic changes as the sole driving forces of a sustainable information society. Projectionistic approaches see superstructures (polity and/or culture) as the determining forces of a sustainable information society. They are the least frequently found approaches in the literature. Dualistic approaches define multiple goals and dimensions of a sustainable information society, but do not consider if these goals are compatible and if and how they are causally linked. Dualistic models are the ones that can be found most frequently in the literature. As an alternative to these three models, the dialectical notion of the participatory, co-operative, sustainable information society (PCSIS) is introduced. Co-operation is based on an inclusive logic that establishes social systems, in which all involved actors benefit. The logic of co-operation is the binding force of a progressive society that connects its various dimensions.

III. Conclusion

Chapter 10: The Digital Commons and the Digital Public Sphere: How to Advance Digital Democracy Today

This chapter asks: what are the democratic potentials of the digital commons and the digital public sphere? First, the chapter identifies ten problems of digital capitalism. Second, it engages with the notion of the digital public sphere. Third, it outlines the concept of the digital commons. Fourth, some conclusions are drawn and ten suggestions for advancing digital democracy are presented.

This chapter contributes to theorising and the analysis of digital capitalism, Internet platforms, the digital public sphere, the digital commons, digital democracy, public service Internet platforms, civil society/community Internet platforms, platform co-operatives, open access, corporate/capitalist open access, and diamond open access.

This work also outlines ten problems of digital capitalism as well as ten principles of digital progressivism, a politics that advances the public sphere and the commons and thereby (digital) democracy in society.

There are natural, economic, political, and cultural dimensions of the commons and the digital commons. Capitalism, public service, and civil society media/community media/co-operatives are three forms of organisation and governing the Internet and digital media/technologies. Capitalism colonises and commodifies the (digital) commons and the (digital) public sphere. Alternative models are located outside of capitalism in the realms of the public sphere and civil society as well as their interactions.

References

Aristotle. 2013a. *Aristotle's Politics*. Translated by Carnes Lord. Chicago, IL: The University of Chicago Press. Second edition.

Aristotle. 2013b. *The Eudemian Ethics of Aristotle*. Translated by Peter L. P. Simpson. New Brunswick, NJ: Transaction.

Aristotle. 2002. *Nicomachean Ethics*. Translated by Joe Sachs. Indianapolis, IN: Hackett.

Aristotle. 1999. *Metaphysics*. Translated by Joe Sachs. Santa Fe, NM: Green Lion Press.

Bentham, Jeremy. 2000. *An Introduction to the Principles of Morals and Legislation*. Kitchener: Batoche Books.

Blackledge, Paul. 2012. *Marxism and Ethics. Freedom, Desire, and Revolution*. Albany: State University of New York Press.

Bynum, Terrell Ward. 2001. Computer Ethics. Its Birth and Its Future. *Ethics and Information Technology* 3 (2): 109–112.

Capurro, Rafael. 2005. Information Ethics. *Computer Society of India Communications*, June 2005: 7–10.

Consalvo, Mia and Charles Ess, eds. 2010. *The Handbook of Internet Studies*. Malden, MA: Wiley-Blackwell.

Dittmer, Joel. 2013. Applied Ethics. In *Internet Encyclopedia of Philosophy*. https://iep.utm.edu/ap-ethic/

Dutton, William, ed. 2013. *The Oxford Handbook of Internet Studies*. Oxford: Oxford University Press.

Engels, Friedrich. 1878. Anti-Dühring. Herr Eugen Dühring's Revolution in Science. In *Marx & Engels Collected Works (MECW)*. *Volume 25*, 5–309. London: Lawrence & Wishart.

Ess, Charles. 2009. *Digital Media Ethics*. Cambridge: Polity Press.

Fieser, James. 2005. Ethics. In *Internet Encyclopedia of Philosophy*. http://www.iep.utm.edu/ethics/

Habermas, Jürgen. 2011. *Zur Verfassung Europas. Ein Essay*. Frankfurt am Main: Suhrkamp.

Habermas, Jürgen. 2008. *Ach, Europa*. Frankfurt am Main: Suhrkamp.

Haines, William. 2006. Consequentialism. In *Internet Encyclopedia of Philosophy*. https://iep.utm.edu/conseque/

Hunsinger, Jeremy, Matthew M. Allen, and Lisbeth Klastrup, eds. 2020. *Second International Handbook of Internet Research*. Dordrecht: Springer.

Hunsinger, Jeremy, Lisbeth Klastrup, and Matthew Allen, eds. 2010. *International Handbook of Internet Research*. Dordrecht: Springer.

Johnson, Deborah G. 2004. Computer Ethics. In *The Blackwell Guide to the Philosophy of Computing and Information*, ed. Luciano Floridi, 65–75. Malden, MA: Blackwell.

Kant, Immanuel. 2011. *Groundworks of the Metaphysics of Morals: A German-English Edition*. Cambridge: Cambridge University Press.

MacIntyre, Alasdair. 1957. What Morality Is Not. *Philosophy* 32 (123): 325–335.

Maner, Walter. 1980. *Starter Kit in Computer Ethics*. Hyde Park, NY: Helvetia Press.

Marx, Karl. 1875. Critique of the Gotha Programme. In *Marx & Engels Collected Works (MECW)*. *Volume 24*, 75–99. London: Lawrence & Wishart.

Marx, Karl. 1844. Contribution to Critique of Hegel's Philosophy of Law. Introduction. In *Marx & Engels Collected Works (MECW)*. *Volume 3*, 175–187. London: Lawrence & Wishart.

McCarthy, George E. 1992, ed. *Marx and Aristotle. Nineteenth-Century German Social Theory and Classical Antiquity*. Savage, MD: Rowman & Littlefield.

McCarthy, George E. 1990. *Marx and the Ancients: Classical Ethics, Social Justice, and Nineteenth-Century Political Economy*. Savage, MD: Rowman & Littlefield.

Meikle, Scott. 1985. *Essentialism in the Thought of Karl Marx*. London: Duckworth.

Moor, James H. 1985. What Is Computer Ethics? *Metaphilosophy* 16 (4): 266–275.

Pike, Jonathan E. 2018. *From Aristotle to Marx. Aristotelianism in Marxist Social Ontology*. London: Routledge.

Spinello, Richard A. 2003. *CyberEthics. Morality and Law in Cyberspace*. Sudbury, MA: Jones and Bartlett.

Tavani, Herman T. 2011. *Ethics and Technology: Controversies, Questions, and Strategies for Ethical Computing*. Hoboken, NJ: John Wiley & Sons. Third edition.

Townsend, Leanne et al. 2016. *Social Media Research: A Guide to Ethics*. http://www.gla.ac.uk/media/media_487729_en.pdf

Chapter Two
Foundations of Communication/Media/Digital (In)Justice

2.1 Introduction

On 3 February 2019, Donald Trump tweeted:

> "With Caravans marching through Mexico and toward our Country, Republicans must be prepared to do whatever is necessary for STRONG Border Security. Dems do nothing. If there is no Wall, there is no Security. Human Trafficking, Drugs and Criminals of all dimensions – KEEP OUT!"[1]

This tweet characterises immigrants from the South as traffickers, drug dealers, and criminals. It makes the sweeping generalisation that immigrants are criminals. Many observers will agree that such a tweet is the communication of ideology and of a particular form of injustice, namely, the reduction of fleeing humans to criminality. The tweet denies immigrants their humanity.

There is a connection between inhumanity and injustice. And in an information society, this connection is frequently communicated in public via the media and the Internet. But what is media/communication (in)justice? And how can we theorise media/communication (in)justice? The chapter at hand is a contribution to the analysis of the media/communication and (in)justice.

Engaging with philosophical approaches to communication, John Durham Peters (1999, 269) concludes that communication is "a political and ethical problem" and that "just communication is an index of the good society". The analysis of communication and society brings up the question of what just communication is and how it should be defined.

DOI: 10.4324/9781003279488-3

This chapter is a contribution to the theoretical debate on media and justice (see also, among others, Christians et al. 2009; Couldry 2012; Habermas 1990; Jansen, Pooley and Taub-Pervizpour 2011; Jensen 2021; Padovani and Calabrese 2014; Rao and Wasserman 2015; Silverstone 2007; Taylor 2017).

Frey et al. (1996, 113) document that until 1996, the journal *Social Justice Research* contained "not a single article written by communication scholars or about communication behavior". *The Philosophical Review*, founded in 1892, published between 1970 and 2020 only seven articles containing the keyword "communication" in the title. *The Journal of Political Philosophy* during the same period just published one article containing "communication" in its title. There is little interest in communication in the field of philosophy. Vice versa, there is also little interest in justice in media and communication studies: between 1970 and 2020, *Journal of Communication* only published nine articles containing the title keyword "justice". In *Communication Theory*, the amount was two articles. There has thus far been a little explicit intersection between ethics/philosophy and communication studies when it comes to the issue of what justice means. This chapter contributes to the intersection of philosophy and media/communication studies for the analysis of (in)justice. It is based on a Marxist-humanist ethics.

In many discussions of ethics, Marxism is either dismissed or not mentioned at all. Marxist theory has something important to add. Unfortunately, it is often not acknowledged as a viable approach to ethics. For example, widely read and cited introductions to ethics and moral philosophy such as McNaughton's (1988) *Moral Vision: An Introduction to Ethics* do not at all mention Marx and Marx-inspired approaches (such as e.g. the ones by Theodor W. Adorno, Paul Blackledge, Gerald A. Cohen, Erich Fromm, Norman Geras, Peter Hudis, Eugene Kamenka, Steven Lukes, [the early, Marxist-humanist works of] Alasdair MacIntyre, George E. McCarthy, Richard W. Miller, Sean Sayers, Michael J. Thompson). The presentation of approaches to ethics is often limited to the discussion of virtue ethics, deontology, and consequentialism (e.g. Fieser 2005).

Also in media ethics and digital ethics, introductions often tend to ignore Marx and the tradition built on his works (e.g. Christians et al. 2017; Floridi 2010, 2013; Patterson, Wilkins and Painter 2019; Ward and Wasserman 2010). My own approach to ethics and critical theory combines, among others, Aristotle's virtue ethics and Marx's critical ethics (Fuchs 2020a). Whereas many Marxists often engage with a variety of non-Marxian approaches, the same cannot be said of many non-Marxian approaches. My point is that Marxian ethics is a legitimate approach that should be more acknowledged in ethics in general as well as in media ethics.

Section 2.2 engages with theories of justice. Section 2.3 outlines foundations of the conceptualising alienation as injustice. Section 2.4 focuses on communication/media (in)justice. Section 2.5 provides a framework for the analysis of digital (in)justice. Some conclusions are drawn in Section 2.6.

2.2 Theories of Justice

Morality is about principles of how to attain the good life and what the difference is between the good and the bad. Ethics is moral philosophy, the systematic theoretical study of morality and morals in society. Justice is one of the key categories in morality and ethics. It has been "a central concern of philosophy from the time of Plato [...] until today" (Rainbolt 2013, 1)

We can distinguish four types of theories of justice (see Table 2.1): idealist monism; dualism; pluralism; and dialectics. The typology is based on logical principles of how to relate two categories: the one and the other. Monism identifies one overarching foundational principle from which others are derived. Dualism identifies two equally foundational and independent substances. Pluralism combines many dualisms so that there are multiple, diverse, independent categories, or principles. It is a special form of dualism. Dialectics is a dialectic of identity and non-identity of the one and the many. There is a unifying principle identical to all aspects and there are interacting, encroaching, intersecting, and diverse moments that have common as well as different aspects.

Idealist justice monism reduces justice to the level of political or cultural justice as key principle. Justice dualism identifies two equally important, independent principles of justice. Justice pluralism is a more complex form of dualism, a combination of several dualisms. It identifies multiple, independent principles of justice. Justice dialectics is based on a unified principle of justice that grounds diverse principles of justice. It is a unity in the diversity of justice that identifies a general principle and grounding of justice

TABLE 2.1   A typology of theories of justice

Logical principle	Theory of justice
Idealist monism	Idealist justice monism
Dualism	Justice dualism
Pluralism	Justice pluralism
Dialectics	Justice dialectics

Theories of Justice

and diverse forms of justice that interact and are based on the unifying principle. It is at the same time monist and pluralist.

2.2.1 Idealist-Monist Theories of Justice

John Rawls's (1999) book *A Theory of Justice* that was first published in 1971 is one of the most cited philosophy books published in the 20th century. In the year 2020, this work had around 90,000 citations,[2] advances a *political monist theory of justice*. It is idealist because its ultimate principle of justice focuses on the realm of politics and political liberties and downplays the importance of the economy. It takes on the form of the *greatest equality principle* that says that all humans have "an equal right to the most extensive total system of equal basic liberties" (Rawls 1999, 266). Rawls gives priority to political justice over socio-economic justice. He argues that the liberal rights that for him make up the constitutional essentials are "more urgent to settle" (Rawls 2001, 49) than social problems. Rawls (1999, 266–267; 2001, 42–43) considers social and economic inequalities as justified as long as the greatest possible benefit is achieved for the least advantaged (*difference principle*) and offices and positions deciding over questions of distribution are open to all (*equal opportunity principle*).

Rawls's theory has been criticised for legitimating class inequality (Cohen 2008; Miller 1975). He characterises liberal rights also as "background justice" (Rawls 2001, 50), which implies a priority of political over socio-economic rights. Although the Universal Declaration of Human Rights defines the right to social security (§22), Rawls basically says that freedom of speech is more important than the right to eat, the right to housing, the right to health, the right to social security, and the right to lead a good life. According to Marxist criticism, Rawls's concept of justice sees it as legitimate that citizens starve as long as they have freedom of speech. Many capitalists will not commit to the difference principle when it implies they have to reduce their profits. They will argue that social policy measures such as higher corporation taxes or the reduction of the working day with full wage compensation destroy their companies and result in unemployment (Miller 1975, 210). The capitalist ruling class has political and ideological institutions at hand that "work exclusively or almost exclusively in its interests" (Miller 1975, 227). It does not voluntarily accept the creation of social equality (Cohen 2008, 290).

Axel Honneth's theory of recognition is a *cultural monist theory of justice* (see Table 2.2). Honneth (2007, 2008) reinterprets Lukács' concept of reification as disrespect. He says that reification is society's disrespect for certain groups and individuals. Disrespect is a

TABLE 2.2 Axel Honneth's theory of recognition

Form of justice	Type of recognition	Institutions
Justice of needs	Love	Family, friendships
Deliberative equality	Equality	Civil society, legal system
Justice of achievement	Achievement, esteem	Solidarity communities of value

lack of recognition for certain human beings in society. Honneth identifies love, equality, and achievement as three forms of respect. For Hegel, these are the recognition of the need for love provided in the family, the recognition of human autonomy in civil society and the legal system, and the recognition of individual particularity by the State and in ethical life and processes of solidarity (Honneth 1996, 25). The absence of such forms or recognition would be the foundation of struggles for recognition. For Honneth, a recon-structive theory of justice needs three normative principles: justice of needs, deliberative equality, and justice of achievement (Honneth 2014, 49).

Honneth does not properly take into account the roles of work, the economy, and use-values in society. The economy seems to simply be another solidarity community pro-viding a particular form of esteem and achievement. The ideal-type economy is about a specific aspect of free human self-realisation through work. In what Marx terms the realm of freedom, work is a source of pleasure, need satisfaction, communication, and care for others. Work is more than a source of achievement. Whereas in *The Struggle for Recognition*, Honneth (1996) tends to ignore the economy, he in *The I in We* (Honneth 2014) subsumes it into the third realm. This means, however, that he reduces the econ-omy and work to recognition. The satisfaction of human needs through social production is primarily a matter of survival and pleasure that cannot be reduced to culture and recognition. Honneth's approach is a "'moral' monism" (Fraser and Honneth 2003, 254), where recognition is the unifying principle of morality and society.

2.2.2 Dualist Theories of Justice

Dualist approaches to justice consider one overarching principle of justice as insufficient and therefore postulate two principles operating in two relatively independent realms of society. Nancy Fraser (1995, 1997) advances a dualist theory of justice (see Young 1997), where the starting point is that "justice today requires *both* redistribution *and* recogni-tion" (Fraser 1995, 69). She considers redistribution and recognition, political economy, and culture/identity, as two relatively distinct and equally important realms of society.

Theories of Justice

TABLE 2.3 Nancy Fraser's theory of justice

Realm of society	Form of justice	Form of injustice
Economy	Distributive justice	Maldistribution
Culture	Recognition	Malrecognition
Politics	Political justice, participatory democracy	Misrepresentation

She discerns between economic and cultural injustices. But in reality, all culture is economic in that it is a realm of the production of meaning (that in contemporary capitalism is often mediated by capital and commodities as the existence of cultural commodities shows) and all economy is cultural because workers have particular working cultures, companies have philosophies, there are corporate ideologies such as neoliberalism, etc. For Fraser, exploitation, economic marginalisation, and deprivation are types of economic injustice, whereas cultural domination, nonrecognition, and disrespect are forms of cultural injustice. Although Fraser acknowledges articulation and interaction, the economy and culture, economic and cultural injustice, and redistribution and recognition are conceptually separate.

In her approach formulated in the 1990s, Fraser held such a two-dimensional concept of justice focused on the economy (distribution) and culture (recognition). Later, she added the concept of political justice (Fraser 2009) and developed her approach into a *pluralistic theory of justice* (see Table 2.3)

Fraser sees the economy, culture, and politics as three equally important and relatively independent domains of society. She argues for a perspectival dualism where the two realms are impinging on each other (Fraser and Honneth 2003, 64). We can characterise Fraser's theory as interactive dualism. For her, the two levels are autonomous and interact in certain cases. Fraser rejects the assumption of a universal normative principle of justice. Recognition, distributive justice, and representation/participation are for her "multiple points of entry into social reality" (Fraser and Honneth 2003, 205). The problem with such an approach is that it establishes a plurality without unity.

2.2.3 Pluralist Theories of Justice

The capabilities approach of Amartya Sen (2009) and Martha Nussbaum (2011) is one of the most influential *pluralistic theories of justice*. Capabilities are about human functioning, being, and doing. In the capabilities approach, justice has to do with the distribution

TABLE 2.4   Iris Marion Young's concept of the five faces of oppression (based on Young 1990, Chapter 2)

Type of oppression
Exploitation
Marginalisation
Powerlessness
Cultural imperialism
Violence

of opportunities for realising capabilities and can be advanced through institutional changes. Nussbaum (2011, 33–34) identifies ten central capabilities: life, bodily health, bodily integrity, senses/imagination/thought, emotions, practical reason, affiliation, concern for other species, play, control over one's political and material environment. The capabilities approach identifies a plurality of human needs, but it remains unclear what unifies and connects human needs.

Iris Marion Young (1990) is another representative of a pluralistic theory of justice. Her theory of (in)justice's key category is oppression. She distinguishes between five forms of oppression that she terms the five faces of oppression (see Table 2.4).

Young rejects the assumption of a human nature and essence:

> "Although social processes of affinity and differentiation produce groups, they do not give groups a substantive essence. There is no common nature that members of a group share. As aspects of a process, moreover, groups are fluid; they come into being and may fade away".
>
> (Young 1990, 47)

The assumption that there are no common features of humans results in incoherent social theory. Humans share capacities such as social production, social and societal relations, self-consciousness, moral reasoning and action, anticipatory thinking, creative action, communication, and co-operation.

A consistent typology has to be complete and its categories have to be non-overlapping. Young's concepts of marginalisation and powerlessness are closely related. For her, powerlessness seems to be marginalisation in the context of decision-making. But such an assumption results in quite narrow concepts of power, empowerment, disempowerment, powerfulness, and powerlessness that are limited to the political dimension of society. Power is the capacity of humans to influence and control their lives and its various

Theories of Justice

dimensions. There is economic, political, and cultural power. Powerlessness is not limited to decision-making but can also take on the form of poverty, voicelessness, invisibility, etc. In disempowerment, humans are robbed of the control of the conditions of their lives. Alienation is the state that results from such disempowerment. Empowerment is the tendency to overcome alienation.

Exclusion and marginalisation are aspects of domination. Domination is defined as the process where one group in society arrives at benefits at the expense of others. The dominating group has advantages and excludes others from such advantages. And it has means at its disposal to defend its privileged position and to keep others disadvantaged. Exclusion is a process through which domination operates. Marginalisation is the result of domination: one group has disadvantages, while another one benefits. Domination operates through a variety of processes and structures, including exclusion, the state, the law, surveillance, violence, warfare, and rules.

Cultural imperialism is one form of disrespect in society. Making other humans, their voices and bodies, invisible through asymmetric power of attention and visibility is one form of disrespect. Scapegoating certain groups is another one. Scapegoating is part of ideology. Ideology is a means and process through which one group portrays society or certain aspects of it (such as certain groups or individuals) in a false or distorted manner in order to legitimate and upholds its power and interests. Cultural imperialism is the privileging of the reputation, visibility, and way of life of one group at the expense of others. It is a unity without diversity that disrespects certain identities and ways of life. But there is also another form of disrespect, namely, diversity without unity, where humans ignore each other and see each other as having nothing in common. Diversity without unity is the imperialism of difference and partiality that ignores commonality and universality. Unity without diversity and diversity without unity are two cultural processes that constitute disrespect. Disrespect is practiced through ideology, by denying other human beings' relevance, or by denying the cultural commons, i.e. common aspects of human life. Young disrespects the complexity of disrespect, especially the oppression caused by difference without unity.

Violence is the intentionally caused physical harm of a human being (Walby 2022). Violence turns the human being "into a thing in the most literal sense: it makes a corpse out of him" (Weil 2005, 183). Violence is not the same as power. It is a dimension of coercive societies and a social relation, where humans try to intentionally cause physical harms to other humans who don't agree to the cause of that harm (see Walby 2022 for a detailed discussion). The harm caused is usually "a physical injury" (Walby et al. 2017,

33), but can also in addition involve mental or psychological harm. Physically injuring others can take on a variety of forms such as assault, torture, rape, killing, murder, war, genocide, enslavement, etc. Violence is a means towards an end such as gaining control of resources (e.g. land, humans), exterminating certain humans, i.e. the absolute exclusion from society through death, gaining pleasure or reputation, etc. Violence is a means for creating alienation, but it is not in itself an alienated system or condition as Young's typology implies.

Intersectional theories are pluralistic theories of justice. Patricia Hill Collins (2000, 299) defines intersectionality as the matrix of domination, whereby she understands

> "the overall organization of hierarchical power relations for any society. Any specific matrix of domination has (1) a particular arrangement of intersecting systems of oppression, e.g., race, social class, gender, sexuality, citizenship status, ethnicity and age; and (2) a particular organization of its domains of power, e.g., structural, disciplinary, hegemonic, and interpersonal".

Figure 2.1 visualises the matrix of domination. It identifies nine plural levels of human identity that are sources of domination. These realms are articulated, but independent.

The problem of pluralistic theories of justice is that they consider society as consisting of independent spheres. Forms of (in)justice are articulated but it remains unclear why there is a particular number of realms and forms of (in)justice. Such categories are often not distinct

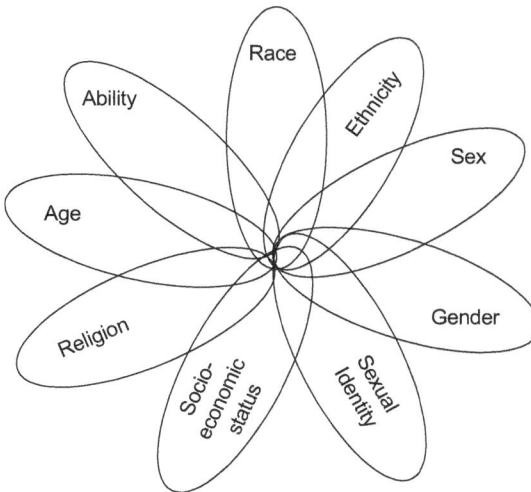

FIGURE 2.1   The matrix of domination in intersectional theories (based on Adams and Zúñiga 2016, 162).

but overlapping, which makes the resulting typologies inconsistent. Eve Mitchell (2013) argues that intersectional theories only stress the difference of identities and lack a focus on the common features of humans and a focus on humanity as the common aspect. According to Mitchell, intersectionality theories often advance relativist theories of (in)justice.

The present author is interested in helping to advance dialectical approaches to studying (in)justice as an alternative to idealist monism, dualism, and pluralism. What follows is the present author's own conception of injustice.

2.3 Alienation as Injustice

The approach taken in this chapter starts from a Marxist-Humanist concept of society (Fuchs 2020a) that stresses human beings' common characteristics and needs in society. Based on Karl Marx, Georg Lukács (1984, 1986) argues that work and social production form the models of human activity in society. He speaks in this context of teleological positing, which is a reference to Aristotle's concept of the teleological cause. Aristotle (2002, §1139b) defines the teleological cause the following way: "one who makes something always makes it for the sake of something". In teleological positing, humans produce active and consciously in order to try to attain defined goals. The dialectic of production and communication is another core common feature of humans: social production is organised through communication and communication is a particular production process, namely, the process of the production of sociality, social relations, social structures, social systems, institutions, and society.

A truly materialist analysis of society does not assume that there is an economic base and a political and cultural superstructure that can be reduced to the base. The materialist analysis of society rather stresses that social production is an economic process that operates in all social relations and social systems, including politics and culture, and takes on emergent qualities in particular systems. There are humans as social producers in the economy, politics, and culture, as well as in all social systems organised in these three realms of society. In the economic system, humans produce use-values that satisfy human needs. In the political systems, they produce collective decisions and rules that govern society's organisation. In culture, they produce meanings and definitions of the world.

A dialectical concept of justice can be based on such a framework of society. It conceives of injustice as alienation and justice as humanism. Alienation is the unifying principle of injustice. Humanism the unifying principle of justice. Alienation and humanism take on different forms in society's various spheres. Marx's notion of alienation is based on the concept of economic alienation but also has a more general meaning. Economic alienation is the class relation, where workers do not own the means of production and the products they are compelled to produce. David Harvey (2018) argues that alienation has a universal character in class societies. The universalisation of alienation is the extension of alienation beyond economic production, the economy, and bounded spaces into realms such as circulation, consumption, culture, politics, globalisation, the relation of nature/society, etc.

Marx sees alienation besides economic exploitation also as the universal form of injustice, in which humans are not in control of the structures that affect their everyday lives (Fuchs 2020b, Chapter 7). Under alienated conditions, humans (re)produce social relations in everyday life and are not in control of the conditions of these social production processes. Alienation is the "production of the object as loss of the object to an alien power, to an *alien* person" (Marx 1844b, 281). Marx characterises alienation in the following words:

> "Under alienated conditions, the human being's "own creation [...] [is] an alien power, his wealth [...] poverty, the *essential bond* linking him with other men [...] an unessential bond, and separation from his fellow men, on the other hand, as his true mode of existence, his life as a sacrifice of his life, the realisation of his nature as making his life unreal, his production as the production of his nullity, his power over an object as the power of the object over him, and he himself, the lord of his creation, as the servant of this creation"".
>
> (Marx 1844a, 217)

Alienation is inhumanity. Alienation implies that humans are robbed of humane living conditions. They are denied parts of their humanity. Given that humans are social beings who depend on each other and produce and communicate in social relations, they all deserve to lead a good life. The need, wish, and desire for a good life are common features of humanity. Alienation is the creation of inhumanity and inhuman conditions. Marx and Engels argue that the "conditions of life of the proletariat sum up all the conditions of life of society today in their most inhuman form" (Marx and Engels 1845, 36–37). The proletariat "cannot abolish the conditions of its own life without abolishing *all* the inhuman conditions of life of society today which are summed up in its own situation" (Marx and Engels 1845, 37). In a class society, there is an "'inhuman' way in which the oppressed class satisfies its needs" (Marx and Engels 1845/1846, 432).

The argument that the ethical foundations of a just society that advances the good life for all as the common good are grounded in human nature as social beings can also be

Alienation as Injustice

found in the communitarian philosophy of Charles Taylor and the Aristotelian philosophy of Alasdair MacIntyre. Taylor stresses that humans are social beings and that advancing justice as the common good follows from this social character: Because

> "of a common good which in fact is sustained by the common life of our society, we ought to accept certain principles of distribution which take account of the real balance of mutual indebtedness relative to this good. For instance, that we owe each other much more equal distribution than we might otherwise agree to on economic criteria, because in face we are involved in a society of mutual respect, or common deliberation, and this is the condition for all of us realizing together an important human potential".

> (Taylor 1985, 298)

MacIntyre (1999) argues that humans are dependent, rational, and communicative animals who depend on each other to survive. "As a practical reasoner, I have to engage in conversation with others, conversation about what it would be best for me or them or us to do here and now, or next week, or next year" (110–111). For achieving the common good, it does not suffice that humans communicate, they also need to co-operate (114). For MacIntyre, the interest to advance the common good for all follows from the social character of humans that makes them depend on each other.

Alienation is a gap between the actuality and the potential of humans and society. They are hindered to be what they could be, to develop to the full extent enabled by society. There are certain parallels between Marx's notion of alienation and the notion of capability development by Sen and Nussbaum. Class societies such as capitalism undermine their own universal promises, there is a

> "*discrepancy between the rhetoric of universal interests and the reality of particular class interests* within the limits circumscribed by particular systems of production and the boundaries of the concomitant social and political institutions and cultural ways of life. The problem can be solved only when the rhetoric/reality discrepancy is overcome, that is, when a particular system of production is established which permits the coincidence of the universal interests of society with the particular interests of a class, and when the concomitant social and political institutions and cultural lifestyles promote and encourage this coincidence. This coincidence results for Marx in the self-realization and self-development of *all* individuals within a society: this outcome has truly become the common good".

> (West 1991, 92)

In a humanist society, all humans and society can realise their potentials and lead a life that is adequate to humans, a humane life. Power differentials in the economy, politics, and culture that privilege the few at the expense of the many have to be overcome for creating a just society. Table 2.5 presents a typology of injustice as alienation.

TABLE 2.5   A typology of injustice as alienation in the economy, politics, and culture

Realm of society	Injustice as alienation	Definition	The alienated	The alienators
Economy	Exploitation	Exploitation is a process where one economic group is capable of and controls means for forcing another economic group to produce goods that are transferred to the dominant class so that it owns and controls these resources.	Exploited class	Exploiters, exploiting class
Politics	Domination	One group in society benefits at the expense of others, who are excluded and marginalised. The dominative group has means at its disposal to defend its privileged position and to keep others disadvantaged. Domination operates through a variety of processes and structures, including exclusion, the state, the law, surveillance, violence, warfare, and rules.	The excluded, marginalised, subalterns	Dictator, dictatorial group
Culture	Disrespect, ideology	When they are disrespected, humans are denied humanity, visibility, attention, recognition through ideology, unity without diversity (imperialism of cultural homogenisation), diversity without unity (cultural relativism, imperialism of cultural difference), cultural asymmetries of voice/visibility/recognition, etc. Ideology is a means and process through which one group portrays society or certain aspects of it (such as certain groups or individuals) in a false or distorted manner in order to legitimate and uphold its power and interests. Disrespect is practiced through ideology, by denying other human beings' relevance, or by denying the cultural commons, i.e. common aspects of human life.	The disrespected	Ideologues, demagogues, influencers

Alienation as Injustice

Alienation is the unifying principle of injustice that takes on specific forms in the economy, the political system, and culture: exploitation in the economy, domination in the political system, and disrespect and ideology in culture.

For Marx, alienation is a feature of capitalism and at the same time older than capitalism. Particular forms of alienation such as war, violence, classes, ideology, or patriarchy are older than capitalism, but have in capitalist society been sublated (*aufgehoben*) in a Hegelian sense: they have been preserved but at the same time transformed into phenomena such as imperialism, the capitalist class, commodity fetishism, reproductive labour that reproduces wage labour, etc. Alienation is not limited to the economy, but connects inequalities across society's different realms. David Harvey (2018) therefore speaks of universal alienation. Alienation is both a condition and a process, a structure and a practice, a state and a relation. In a dialectical process, alienation interconnects the levels of objects and human subjects in societies that are shaped by domination.

The capitalist economy is a system, in which workers produce commodities with the help of means of production that are the private property of companies. These commodities are sold on commodity markets so that profit is achieved and capital can be accumulated. In a capitalist society, the logic of accumulation also extends into the political and cultural system where we find the accumulation of decision-power and influence in the political system and the accumulation of reputation, attention, and respect in the cultural system. Table 2.6 gives an overview of injustices in capitalist society. The accumulation of capital, influence, and reputation results in the asymmetrical distribution of economic, political and cultural power and creates the injustices of exploitation of labour in the capitalist economy, domination of citizens in the political system, and disrespect of human individuals and groups in the cultural system.

Based on the concept of injustice as alienation, we will next discuss the relation of communication and injustice.

2.4 Communication/Media Injustice

Aristotle stresses that justice has to do with the common good, the common benefit for all. "And the proverb 'the things of friends are common' is right, since friendship consist in community. All things are common to brothers and comrades" (Aristotle 2002, §1159b). In a just society, all humans are friends and are enabled to treat each other as friends. There are joint etymological roots of the words "communication", "community", and the "commons". To "communicate" meant originally to make something "common

TABLE 2.6   Alienation as injustice in capitalist society

Sphere	General features	Structure	Process	Antagonism	Injustice
Economy	Production of use-values	Class relation between capital and labour	Capital accumulation	Capital vs. labour	Capitalist exploitation: capital's private ownership of the means of production, capital, and created products implies the working class' non-ownership and exploitation
Politics	Production of collective decisions	Nation-state	Accumulation of decision-power and influence	Bureaucracy vs. Citizens	Domination: citizens' lack of influence on political decisions as consequence of the asymmetric distribution of power
Culture	Production of meanings	Ideologies	Accumulation of reputation, attention and respect	Ideologues and celebrities vs. everyday people	Invisibility, disrespect: lack of recognition as consequence of an asymmetric attention economy

to many" (Williams 1983, 72). A true communication society is a society of the commons where everyone benefits (Fuchs 2020a). But communication has just like society taken on alienated forms.

For both Nancy Fraser and Iris Marion Young, communication is a cultural phenomenon (see Fraser 1997, 13–14; Young 1990, 23, 38). They leave open the relationship of communication and work. One can clearly see the influence of Habermas on Fraser's and Young's approaches. Habermas separates work and interaction, which resulted in his dualism of system and lifeworld (Fuchs 2020a). The basic problem of this dualism is that communication is not limited to a specific realm of society but is together with production constitutive of all social relations. Production is communicative just like communication is a specific production process. The dialectic of communication and production shapes all realms of society. Communication is therefore not limited to culture.

As a particular type of production and teleological positing that (re)produces sociality and social relations, communication is an inherent feature of the social relations of humans in society. Alienation therefore always has a communicative dimension. Table 2.7 provides an overview of communication's roles in alienation. In class relations, humans are compelled to act and communicate in order to produce goods that are owned by the ruling class. In alienated politics, humans are excluded from influential political communication

Communication/Media Injustice

TABLE 2.7   The communicative dimension of different forms of injustice

Type of injustice	Communicative dimension
Economic alienation: exploitation	Class communication: communication in class relations that organise exploitation
Political alienation: domination	Exclusion from political communication
Cultural alienation: disrespect, ideology	Invisibility of voice:
	what certain individuals and groups say and do is presented in distorted manners or is marginalised

that makes a political difference (exclusion), or their voices are marginalised (marginal-isation), or their communication and information are monitored (surveillance), or their minds and bodies, including their communication, are absolutely silenced through geno-cide, murder, war, etc. (violence). In alienated culture, there are asymmetries of reputa-tion, which means that culturally marginalised individuals and groups might be able to speak, but they are not heard, or they are hardly or seldomly heard, or what they say and do is through ideologies presented in distorted ways in the public so that their reputation is harmed and what they think, say, and do is perceived in false ways.

Communication is the process where two or more humans interact symbolically in order to make meaning of each other and the world. Media are means of communication, means that mediate, i.e. enable and support, communication. Media include, for exam-ple, sound and light in the case of face-to-face communication and media technologies such as the book (print media), radio and television (audio–visual media), the telephone and the Internet (interactive media). Communication is a human practice. Media are me-diating structures. There are media wherever there is the interaction of moments. For example, the blood system and the brain are mediating systems of the body. In society, media are means of communication. Wherever humans communicate, there is some form of mediation. Wherever there are media in society, there are human information and communication processes. There is a dialectic of communication and the media. In alienated societies, media and communication tend to take on alienated forms. Just like communication is an aspect of alienation, alienation is also an aspect of communication. In alienated societies, there is a dialectic of alienation and communication. There is communicative alienation and alienated communication. Table 2.8 provides an overview of the dimensions of alienated communication and alienated media.

In the communication and media economy, alienation is the exploitation of communica-tion workers and private control of means of communication so that others are excluded

TABLE 2.8   Forms of alienated communication and alienated media/means of communications

Dimension of injustice	Alienated communication	Alienated media/means of communication
Economic injustice	Exploitation of communication workers; humans are economically disabled from or limited in producing, disseminating, or consuming information	Private ownership of the means of communication
Political injustice	Exclusion of humans and their voices from political communication that influences political decisions	Dictatorial governance of the means of communication
Cultural injustice	The production and dissemination of ideology and the (re)production of asymmetries of attention and visibility of communication	Ideological means of communication that advance malrecognition

TABLE 2.9   The interaction of class, racism, gender oppression

	Class	Racism	Gender-related oppression, patriarchy
Class	Exploitation	Racist exploitation	Gender-structured exploitation
Racism	Racist exploitation	Racism	Discrimination of racialised individuals or groups of a particular gender
Gender-related oppression, patriarchy	Gender-structured exploitation	Discrimination of racialised individuals or groups of a particular gender	Gender-based discrimination

from owning and using these means of production. In communication and media politics, alienation is the exclusion of certain groups' and individuals' voices from influential political communication and the existence of dictatorial decision-making processes in media organisations. In communication and media culture, alienation is the production and dissemination of ideology via means of communication and the production and reproduction of asymmetries of attention and visibility. There is an asymmetric attention economy.

Capitalism, racism, and patriarchy are three modes of power relations that each combine economic alienation, political alienation, and cultural alienation. Capitalism, racism, and patriarchy involve specific forms of exploitation, domination, and ideology. The three forms of alienation are interacting in particular forms of power relations. Capitalism, racism, and patriarchy/gender-related oppression are inherently connected and interacting. The economy plays a particular role in this interaction because these power relations are relations of production and accumulation of power. Table 2.9 provides an overview of the interactions of capitalism, racism, and patriarchy.

The capitalist economy creates forms of highly exploited, insecure, precarious labour, including racialised labour, unpaid labour, reproductive labour, and gender-defined labour, in order to maximise profits. Racism and patriarchy have economic, political, and ideological dimensions. In capitalism, these dimensions are united by the logic of accumulation. Class, racism, and gender oppression/patriarchy are the three main forms of power relations that advance alienation, deny humans their humanity, and create damaged lives.

The interaction of class, racism, and gender oppression matters in the context of communication and media injustices. Any intersection of these power systems has communicative features and shapes communication(s) in societies structured by exploitation and domination. For example, the intersection of capitalism and racism in the context of communication and the media takes on the form of the super-exploitation of communication workers (e.g. call centre agents) of colour and immigrant communication workers, who are forced to work for lower wages than others and are the first to be fired.

In the next section, we will focus on a particular type of communication/media injustice, namely, digital injustice.

2.5 Digital Injustice

Digital communication and digital media are particular forms of communication and the media. A central feature of digital technologies is that the computer and digitalisation enable the convergence of information-, communication- and production-technologies. The computer is not just a digital communication medium, but also a digital machine, i.e. an instrument of work and a technology of co-operation. The networked computer enables the prosumption of information, where consumers and users are enabled to produce user-generated content.

Table 2.10 gives an overview of three forms of alienation in digital society, i.e. of forms of digital injustice.

Digital exploitation and digital destructive forces constitute economic forms of digital alienation. Digital capital's exploitation of digital labour plays an important role in economic digital alienation. But digital production can also have negative and destructive effects on nature and the health of human beings. In such cases, the digital productive forces become digital destructive forces. In the realm of digital politics, alienation takes on the form of digital domination: digital technologies are used as means of dictatorship,

TABLE 2.10 Types of digital alienation

Dimension of society	Form of digital alienation	Antagonism	Example
Digital economy	Digital exploitation, digital destructive forces	Digital capital/digital labour	Exploitation of digital workers such as Foxconn assemblers or Uber's platform workers; the monopoly power of Google, Facebook, Apple, Amazon, Microsoft, etc.; the destructive effects of digital technologies on nature and humans (e.g. the poisoning of soil, water and humans by e-waste)
Digital politics	Digital domination	Digital dictators/digital citizens	Donald Trump's use of Twitter and other social media; dictatorial regime's digital surveillance of citizens; digital warfare
Digital culture	Digital ideology, digital disrespect	Digital ideologues and influencers/ digital human beings: asymmetrical attention economy, ideology on and about the Internet	Popular culture on social media: the cultural power of online-influencers such as PewDiePie (>100 million followers on YouTube); the communication of racist and nationalist ideology on the Internet

surveillance, exclusion, control, war, and violence. In digital culture, alienation is digital ideology and digital disrespect: ideologies such as online nationalism or online racism are spread via digital networks; humans are disrespected in Internet communication, for example by asymmetries of online voice, attention, and visibility; there are ideologies about the Internet (such as digital techno-determinism, digital techno-optimism, and digital techno-pessimism).

2.6 Conclusion

In theorising justice, dialectical theories of justice are alternatives to idealist monism, dualism, and pluralism. The approach outlined in the chapter at hand is a contribution to dialectical concepts of justice, communication justice, media justice, and digital justice. It utilises the theoretical approach of Marxist Humanism. Marxist Humanism argues that there is an essence of human beings, i.e. common features such as social production and the dialectic of communication and production. Alienation is a key concept of Marxist Humanism. Using this approach, injustice can be characterised as alienation. Alienation is the common aspect of injustice. Alienation destroys the realisation of the potentials of humans and society. It creates power asymmetries and a gap between potentiality and actuality in society.

Injustice as alienation takes on economic, political, and cultural forms in society in unjust, alienated societies in general and in communication processes, media organisations/

systems, and in the context of digital technologies that are part of alienated societies. Alienation creates inhumanity. It denies human beings their humanity by limiting their human capacities and the realisation of society's potentials.

In order to overcome alienation, social struggles for a just society and just societal conditions are needed. Progressive social movements are the practical dimension and expression of struggles for justice and protests against injustices. Protest movements utilise means of communication for protest organisation and public mobilisation. And there are also movements for communicative justice that make political demands to change the conditions of communication in society in order for trying to advance democratic, humanist communication and democratic, humanist means of communication. Humanism is the negation of the negation of alienation. Humanist communication and means of communication are the negation of the negation of alienated communication(s). Table≈2.11 provides an overview of the humanist organisation of society's various realms.

Table 2.12 provides an overview of forms of communication and media justice.

Humanist, just communication is socialist media/communication in the economy, democratic media/communication in politics, and respectful communication and media that are a source of the recognition of everyone. Humanist, just communication stands in an antagonism to class-based, exploitative media/communication, dictatorial media/communication, and ideological media/communication that advance malrecognition and asymmetries of voice.

TABLE 2.11 Dimensions of a humanist society

Realm of society	Forms of justice	Meaning of justice
Economy: economic justice	Socialism	Self-managed economic organisations where the means of production are collectively owned and controlled, wealth for and self-realisation of all humans
Politics: political justice	Participatory democracy	All humans are enabled to participate in the decision-making processes that concern their lives
Culture: cultural justice	Respect, recognition	Human beings and groups are welcomed and their interests, identities, worldviews, and lifestyles are recognised; there is unity in diversity of identities, worldviews, interests, and lifestyles

TABLE 2.12   Forms of humanist and just communication/media

Dimension of justice	Humanist, just communication	Humanist, just media/means of communication
Economic justice	Socialist communication: worker self-management of communication companies; enablement of humans to produce, disseminate, and consume information;	Socialist media: collective ownership of the means of communication (public service media, citizen media); information and information technologies as common and public goods
Political justice	Democratic communication:participation of humans in political communication so that their voices are heard and make a collective difference	Democratic media: democratic governance of the means of communication
Cultural justice	Respectful communication: the production and dissemination of respect and an inclusive culture that enables everyone to be visible in the public sphere; unity in diversity of voices; education in how to argue in complex and intelligent ways and make one's critical voice heard; respectful, complex, controversial, critical debate and constructive disagreement	Media of recognition: friendly and inclusive means of communication that make humans' interests and voices heard and respected by others

Table 2.13 provides an overview of just digital communication(s).

Contemporary societies are capitalist societies. Capitalism signifies the negativity of accumulation: the existence of injustices. Communicative and digital capitalism are unjust societies with large power asymmetries. The alternative is a just, humanist communicative and digital society of the commons, where communication's original meaning as making something common is realised so that all humans benefit. Attaining a true communication society requires first and foremost praxis, i.e. social struggles against the injustices of alienation, namely, exploitation, domination, ideology/disrespect/malrecognition.

For Marxist Humanism, justice, communication justice, media justice, and digital justice are not abstract ideas. Ethics and justice are only material and humanist if they are not limited to the realm of concepts and interpretation, but take on the form of praxis in social movements. Communication justice has to be part of broader struggles for a democratic society of the commons, a participatory democracy.

Conclusion

TABLE 2.13   A typology of just, humanist digital communication(s)

Realm of society	Type of humanist, digital justice	Meaning of humanist, digital justice
Economy	Digital socialism	Network access for everyone, community is in control of technology, digital resources as common goods, green computing/ICTs
Politics	Digital democracy: participation and democracy in decision-making	Digital technologies support participatory and deliberative democracy and inclusive political communication in the public sphere
Culture	Digital recognition	Digital media/communication support making the voices of all heard, recognition of all; the unity of diversity of identities, lifestyles and worldviews; education in obtaining digital skills that help practicing unity in diversity socialism, and democracy

Notes

1 https://twitter.com/realdonaldtrump/status/1092181733825490945.
2 Data source: https://scholar.google.com/scholar?cites=51401311553152067 11&as_sdt=2005 &sciodt=0,5&hl=en, accessed on 1 September 2020.

References

Adams, Maurianne and Ximena Zúñiga. 2016. Getting Started: Core Concepts for Social Justice Education. In *Teaching for Diversity and Social Justice*, ed. Maurianne Adams and Lee Anne Bell with Diana J. Goodman and Khyati Y. Joshi, 143–184. New York: Routledge. Third edition.

Aristotle. 2002. *Nicomachean Ethics*. Indianapolis, IN: Hackett.

Christians, Clifford G., Mark Fackler, Kathy Brittain Richardson, Peggy J. Kreshel, and Robert H. Woods Jr. 2017. *Media Ethics: Cases and Moral Reasoning*. New York: Routledge. Tenth edition.

Christians, Clifford G., Theodore L. Glasser, Denis McQuail, Kaarle Nordenstreng, and Robert A. White. 2009. *Normative Theories of the Media. Journalism in Democratic Societies*. Chicago: University of Illinois Press.

Cohen, Gerald A. 2008. *Rescuing Justice & Equality*. Cambridge, MA: Harvard University Press.

Collins, Patricia Hill. 2000. *Black Feminist Thought: Knowledge, Consciousness, and the Politics Empowerment*. New York: Routledge. Second edition.

Couldry, Nick. 2012. *Media, Society, World. Social Theory and Digital Media Practice*. Cambridge: Polity.

Fieser, James. 2005. Ethics. In *Internet Encyclopedia of Philosophy*. http://www.iep.utm.edu/ethics/

Floridi, Luciano. 2013. *The Ethics of Information*. Oxford: Oxford University Press.

Floridi, Luciano, ed. 2010. *The Cambridge Handbook of Information and Computer Ethics*. Cambridge: Cambridge University Press.

Fraser, Nancy. 2009. *Scales of Justice. Reimagining Political Space in a Globalizing World*. New York: Columbia University Press.

Fraser, Nancy. 1997. *Justice Interrupts. Critical Reflections on the "Postsocialist" Condition*. New York: Routledge.

Fraser, Nancy. 1995. From Redistribution to Recognition? Dilemmas of Justice in a 'Post-Socialist' Age. *New Left Review* I/212: 68–93.

Fraser, Nancy and Axel Honneth. 2003. *Redistribution or Recognition? A Political-Philosophical Exchange*. London: Verso.

Frey, Lawrence R. et al. 1996. Looking for Justice in All the Wrong Places: On a Communication Approach to Social Justice. *Communication Studies* 47 (1–2): 110–127.

Fuchs, Christian. 2020a. *Communication and Capitalism: A Critical Theory*. London: University of Westminster Press. https://doi.org/10.16997/book45

Fuchs, Christian. 2020b. *Marxism. Karl Marx's Fifteen Key Concepts for Cultural and Communication Studies*. New York: Routledge.

Habermas, Jürgen. 1990. *Moral Consciousness and Communicative Action*. Cambridge, MA: The MIT Press.

Harvey, David. 2018. Universal Alienation. *tripleC: Communication, Capitalism & Critique* 16 (2): 424–439. https://doi.org/10.31269/triplec.v16i2.1026

Honneth, Axel. 2014. *The I in We. Studies in the Theory of Recognition*. Cambridge: Polity.

Honneth, Axel. 2008. *Reification. A New Look at an Old Idea*. Oxford: Oxford University Press.

Honneth, Axel. 2007. *Disrespect: The Normative Foundations of Critical Theory*. Cambridge: Polity.

Honneth, Axel. 1996. *The Struggle for Recognition. The Moral Grammar of Social Conflicts*. Cambridge: Polity.

Jansen, Sue Curry, Jefferson Pooley, and Lora Taub-Pervizpour, eds. 2011. *Media and Social Justice*. Basingstoke: Palgrave Macmillan.

Jensen, Klaus Bruhn. 2021. *A Theory of Communication and Justice*. London: Routledge.

Lukács, Georg. 1986. *Zur Ontologie des gesellschaftlichen Seins. Zweiter Halbband. Georg Lukács Werke, Band 14*. Darmstadt: Luchterhand.

Lukács, Georg. 1984. *Zur Ontologie des gesellschaftlichen Seins. Erster Halbband. Georg Lukács Werke, Band 13*. Darmstadt: Luchterhand.

MacIntyre, Alasdair. 1999. *Dependent Rational Animals. Why Human Beings Need the Virtues*. Chicago, IL: Open Court.

Marx, Karl. 1844a. Comments on James Mill, Élémens d'economie politique. In *Marx & Engels Collected Works (MECW) Volume* 3, 211–228. London: Lawrence & Wishart.

Marx, Karl. 1844b. Economic and Philosophic Manuscripts of 1844. In *Marx & Engels Collected Works (MECW) Volume 3*, 229–346. London: Lawrence & Wishart.

Marx, Karl and Friedrich Engels. 1845/1846. The German Ideology. Critique of Modern German Philosophy According to Its Representatives Feuerbach, B. Bauer and Stirner, and of German Socialism According to Its Various Prophets. In *Marx & Engels Collected Works (MECW) Volume* 5, 15–539. London: Lawrence & Wishart.

Marx, Karl and Friedrich Engels. 1845. The Holy Family, or Critique of Critical Criticism. Against Bruno Bauer and Company. In *Marx & Engels Collected Works (MECW) Volume 4*, 9–211. London: Lawrence & Wishart.

McNaughton, David. 1988. *Moral Vision: An Introduction to Ethics*. Malden, MA: Blackwell.

Miller, Richard W. 1975. Rawls and Marxism. In *Reading Rawls. Critical Studies on Rawls' "A Theory of Justice"*, ed. Norman Daniels, 206–230. New York: Basic Books.

Mitchell, Eve. 2013. I Am a Woman and a Human: A Marxist-Feminist Critique of intersectionality Theory. https://libcom.org/files/intersectionality-pamphlet.pdf (accessed on 1 September 2020).

Nussbaum, Martha. 2011. *Creating Capabilities. The Human Development Approach*. Cambridge, MA: The Belknap Press.

Padovani, Claudia and Andrew Calabrese, eds. 2014. *Communication Rights and Social Justice: Historical Accounts of Transnational Mobilizations*. Basingstoke: Palgrave Macmillan.

Patterson, Philip, Lee Wilkins, and Chad Painter, eds. 2019. *Media Ethics. Issues and Cases*. Lanham, MD: Rowman & Littlefield.

Peters, John Durham. 1999. *Speaking into the Air: A History of the Idea of Communication*. Chicago, IL: The University of Chicago Press.

Rao, Shakuntalao and Herman Wasserman, eds. 2015. *Media Ethics and Justice in the Age of Globalization*. Basingstoke: Palgrave Macmillan.

Rainbolt, George W. 2013. Justice. In *International Encyclopedia of Ethics*, ed. Hugh LaFollette. Malden, MA: Blackwell. https://doi.org/10.1002/9781444367072.wbiee385

Rawls, John. 2001. *Justice as Fairness. A Restatement*. Cambridge, MA: Belknap Press.

Rawls, John. 1999. *A Theory of Justice*. Cambridge, MA: Belknap Press. Revised edition.

Sen, Amartya. 2009. *The Idea of Justice*. Cambridge, MA: The Belknap Press.

Silverstone, Roger. 2007. *Media and Morality: On the Rise of the Mediapolis*. Cambridge: Polity.

Taylor, Charles. 1985. *Philosophy and the Human Sciences. Philosophical Papers 2*. Cambridge: Cambridge University Press.

Taylor, Linnet. 2017. What Is Data Justice? The Case for Connecting Digital Rights and Freedom Globally. *Big Data & Society* 4 (2). https://doi.org/10.1177%2F2053951717736335

Walby, Sylvia. 2022. *Theorizing Violence*. Cambridge: Polity.

Walby, Sylvia et al. 2017. *The Concept and Measurement of Violence against Women and Men*. Bristol: Policy Press.

Ward, Stephen J. A. and Herman Wasserman, eds. 2010. *Media Ethics beyond Borders: A Global Perspective*. New York: Routledge.

Weil, Simone. 2005. *An Anthology*. London: Penguin.

West, Cornel. 1991. *The Ethical Dimensions of Marxist Thought*. New York: Monthly Review Press.

Williams, Raymond. 1983. *Keywords. A Vocabulary of Culture and Society*. New York: Oxford University Press. Revised edition.

Young, Iris Marion. 1997. Unruly Categories: A Critique of Nancy Fraser's Dual Systems Theory. *New Left Review* I/222: 147–160.

Young, Iris Marion. 1990. *Justice and the Politics of Difference*. Princeton, NJ: Princeton University Press.

Chapter Three
The Ethics of the Digital Commons

3.1 Introduction

The rise of the computing and the Internet in society has come along with new forms of commodities and commons. There is a range of digital commodities: Apple sells hardware such as iPhones, iPads, or Macintosh desktop computers. Internet service providers sell access to digital networks for mobile phones, laptops, and desktop computers. Microsoft sells licenses for the use of operating system software and application software. Google and Facebook sell targeted online advertisements. Spotify, Netflix, and Amazon Prime sell subscriptions to collections of digital content.

There is also a range of digital commons: community centres, public libraries, and other public institutions often provide gratis access to computers and the Internet. Community networks such as Freifunk provide gratis access to computer networks that are operated and owned as a common resource in local communities. Free software (such as Linux, GNU, or Mozilla) is software that can be run, studied, distributed, and improved without restrictions. Wikipedia is a freely accessible, co-operatively edited, non-profit online encyclopaedia that volunteers edited co-operatively. It is distributed based on a Creative Commons licence that allows re-use and re-mixing of Wikipedia's content. Creative Commons is a licence that allows access to and the re-use of digital contents (such as images, texts, videos, and music) without payment. Non-profit open access journals and books make texts available in digital online formats (and in the case of books often as affordable paperbacks) without charging users and authors and without making monetary profits.

DOI: 10.4324/9781003279488-4

Digitisation fosters both new forms of commodification and co-operative production, distribution and ownership, which poses the ethical question of how we should best assess these diverging principles that operate in the online economy. *This chapter asks: Why is it morally good to foster the digital commons? It studies ethical foundations of the commons and applies them to the realm of the digital commons. This task falls into the realm of computer ethics, which is a subdomain of ethics. The focus of this chapter is on a particular version of ethics, namely Aristotelian ethics in general and Alasdair MacIntyre's version of Aristotelian ethics in particular. The chapter explores how this approach can be interpreted for justifying the moral need for digital commons. It outlines foundations of an Aristotelian-Hegelian-Marxian digital ethics.*

The rise of the Internet has advanced new forms of commodities and commons. Digital commodities Hess and Ostrom (2003) define a common good as a resource that features high subtractability and the difficulty of excluding others from access and use. High subtractability is also known as rivalrous consumption. So Hess and Ostrom base their understanding of the commons on a theory of economic goods (see Hess and Ostrom 2007, 9 and Table 3.1). Hess and Ostrom (2003, 120) decouple the concept of the commons from the question of ownership. They argue that common-pool resources can be owned by a government, a community, corporations, or private individuals.

Yochai Benkler in contrast to Ostrom defines the commons in opposition to markets:

The commons are "radically decentralized, collaborative, and nonproprietary; based on sharing resources and outputs among widely distributed, loosely connected individuals who cooperate with each other without relying on either market signals or managerial commands. This is what I call 'commons-based peer production'" (Benkler 2006, 60).

TABLE 3.1 Types of goods in the economic theory of goods

		Subtractability	
		Low	**High**
Exclusion	**Difficult**	**Public goods**	**Common-pool resources**
		Useful knowledge, Sunsets	Libraries, Irrigation systems
	Easy	**Toll or club goods**	**Private goods**
		Journal subscriptions, Day-care centres	Personal computers, Doughnuts

> "Commons are an alternative form of institutional space, where human agents can act free of the particular constraints required for markets, and where they have some degree of confidence that the resources they need for their plans will be available to them. Both freedom of action and security of resource availability are achieved in very different patterns than they are in property-based markets".
>
> (Benkler 2006, 144)

Benkler (2006, 61–62) discerns four types of commons, for which he uses two criteria: (1) first, the question is whether there is open access to the commons or access only for a particular community. (2) Second, the question is whether or not there is legal regulation of the use and access to the commons.

Ostrom and Benkler are probably the two most widely read 21st-century scholars who have published on the commons. Their major books *Governing the Commons* (Ostrom 1990) and *The Wealth of Networks* (Benkler 2006) are political theories of the commons that ask: how can the commons be governed? The task of Ostrom's (1990, 27) book is to "develop a series of reasoned conjectures about how it is possible that some individuals organize themselves to govern and manage CPRs [common-pool resources] and others do not". Benkler's book (2006) asks how the digital commons – commons at the level of digital infrastructures, software technologies and digital culture/content – should be governed. He asks: "To what extent will resources necessary for information production and exchange be governed as a commons, free for all to use and biased in their availability in favor of none?" (Benkler 2006, 23).

In neither of the two books can one find a philosophical engagement with the question: why do humans and society need the commons? Both approaches leave out ethical questions. This chapter in contrast explores ethical foundations of how to justify the need for digital commons.

Section 3.2 analyses how Alasdair MacIntyre's version of Aristotelian ethics relates to the digital commons. Section 3.3 focuses on the notion of human essence. Section 3.4 gives attention to the concept of the common good. Section 3.5 brings together the preceding discussions by using the notions of human essence and the common good as the foundation for morally justifying the digital commons. Section 3.6 discusses some concrete examples that apply Aristotelian–Hegelian–Marxian digital ethics. Section 3.7 draws some conclusions.

Introduction

3.2 Alasdair MacIntyre and the Digital Commons

Although Yochai Benkler does not in general give much attention to ethics and philosophy in his books on the commons, a paper he wrote together with Helen Nissenbaum discusses "Commons-Based Peer Production and Virtue" (Benkler and Nissenbaum 2006).

Benkler and Nissenbaum (2006, 414) argue that "commons-based peer production is an instance of an activity that not only enables the expression of virtuous character but serves as a training ground for virtue" and "holds the potential to add to the stock of opportunities for pro-social engagement". They use virtue ethics for discerning four clusters of virtues that motivate commons-based peer production. The first two clusters focus on the development of the commoners' self (self-regarding virtues), the third and the fourth on the development of others (social virtues) (Benkler and Nissenbaum 2006, 405–408):

- Virtue cluster 1: Autonomy
- Virtue cluster 2: Creative production
- Virtue cluster 3: Benevolence, charity, generosity, altruism
- Virtue cluster 4: Sociability, camaraderie, friendship, co-operation, civic virtue
- Virtue cluster 1: Autonomy "[P]articipation in peer production constitutes an arena of autonomy, an arena where they [the commoners] are free to act according to self-articulated goals and principles" (Benkler and Nissenbaum 2006, 405)
- Virtue cluster 2: Creative production "[P]eer production opens up new avenues for creative, productive practices. [...] peer production offers a medium for contributing our thoughts, our knowledge, our know-how [...] to a meaningful product" (406).
- Virtue cluster 3: Benevolence, charity, generosity, altruism Peer production is often motivated by "the pleasure or satisfaction of giving – generosity, kindness, benevolence" (408).
- Virtue cluster 4: Sociability, camaraderie, friendship, co-operation, civic virtue Peer production means the virtue of giving "to a commons, a community, a public, a mission, or a fellowship of which the giver is a part", "to be part of a collective effort", and "to give or produce something of value to all" (408).

In the discussion of virtue cluster 2, Benkler and Nissenbaum (2006, 406) refer to Alasdair MacIntyre's version of virtue ethics: "Peer production offers the possibility of engagement in what MacIntyre terms a 'practice'". They then cite MacIntyre's (2007, 187) definition of practice as "any coherent and complex form of socially established human activity through which goods internal to that form of activity are realized in the course of trying to achieve those standards of excellence which are appropriate to, and partially derivative of, that form of activity, with the result that human powers to achieve excellence, and human conceptions of the ends and goods involved, are systematically extended".

However, Benkler and Nissenbaum do not further discuss the implications of this definition.

A good is "what benefits human beings as such and [...] what benefits human beings in particular roles within particular contexts of practice" (MacIntyre 1999, 65). There are therefore individual and social goods. MacIntyre (2007, 291) mentions prestige, status, and money as examples of external goods. Internal goods arise directly from the experience of a practice itself (292). MacIntyre relates the virtue concept to practices' internal goods: "*A virtue is an acquired human quality the possession and exercise of which tends to enable us to achieve those goods which are internal to practices and the lack of which effectively prevents us from achieving any such goods*" (222). For Aristotle (2002, book II), a virtue is an active condition that constitutes a mean between two extremes – "a mean condition between two vices, one resulting from excess and the other from deficiency" (2002, §1107a). So, a virtue is a means of moderation and mediation. Theodor W. Adorno (2001) criticises that Aristotle's ethics is conservative (see also Whyman 2017). For Aristotle, mediation is "only something existing *between* the extremes" (Adorno 2001, 47). This concept of mediation lacks the dialectic, in which mediation is "accomplished *through* the extremes themselves" (47) and the extremes are sublated. So, for example, class struggle aims at sublating the conflict between capital and labour that is constituted by two opposite interests. A "mediation" of this conflict is achieved by mechanisms such as wage negotiations, strikes, lay-offs, rationalisation, outsourcing, etc. Within capitalism, such mediation only settles the conflict temporarily closer to the interest of capital or labour and cannot overcome the extreme polar opposite of interests that is constitutive for capitalism itself. Sublation (*Aufhebung*) works through the extremes by constituting a new totality that is an emergence difference that contains the new and transformed parts of the old, eliminates parts of the old, and makes a qualitative difference. The conservative character of Aristotle's ethics can only be overcome by a focus on ethico-political of social struggle that aims at sublating the social contradictions that constitute exploitation and domination.

Alasdair MacIntyre and the Digital Commons

The problem of defining virtues as individual practices is that one can obtain internal excellence of a practice in order to perfect ways of achieving external goals that harm society. An example is someone who acquires outstanding free/open-source software programming skills, uses them for building bots that tyrannise specific groups of Internet users, and encourages others to re-use and further develop the evil code in order to create an army of nasty bots. Virtues therefore need to be situated in the context of the political and critical dimension of the quest and struggle for a good life (εὐδαιμονία, eudaimonia) for all.

Creative production is not, as Benkler and Nissenbaum claim, a virtue in itself. It should not be seen independent from its content and societal context. MacIntyre (2007) takes this point into account by arguing that virtues do not simply focus on the establishment of the good life for an individual, but a community. Neo-Aristotelians ask: "What is that *we* presuppose? [...] the NeoAristotelian's history is a history both of her and of those groups with whom she shares common goods and within which she pursues her individual good" (MacIntyre 2016, 61). But such a version of the communitarian approach is itself limited: The Nazis in Nazi-Germany had the virtue of perfecting their methods of annihilating their constructed enemies (such as Jews), which was perceived as creating excellence in militarism and a good life for the community of Nazis. The only community at which virtues can be oriented in order not to be repressive is the undivided community of humans. Practices can only be virtuous if they aim at creating a good life for all and to reduce and minimise suffering. Callinicos (2011, 77) remarks that MacIntyre has a "principled preference for the local and particular" that has to do with the circumstance that "MacIntyre has lost interest in the search for a global alternative (literally and metaphorically) to capitalism". MacIntyre (2011) answers to Callinicos that revolutions need to start from organisations such as "grass-roots organisations, trade unions, cooperatives", schools, transport systems, etc. that help remaking everyday life and "serve the common good" (320). "For those who engage in such making and remaking will encounter that resistance to any breach of those [dominant political] norms" is what "makes revolutionaries", a resistance for which local organisations need "to find allies elsewhere, nationally and internationally, and often need to deal with agencies of the state or international agencies, sometimes as obstacles, sometimes as providing resources" (320).

Floridi (2013, 164) argues that virtue ethics "is intrinsically *egopoietic*" and can therefore turn into "ethical individualism" (167). It faces the problem of scaling up from the individual and local communities to "global values" (168) and the "globalized world in general and [...] the information society in particular" (167). The problem of Floridi's approach, however, is that it scales up beyond human values and society and advances a moral relativism that equalises humans and non-humans (Fuchs 2016). Ess and Fossheim (2013) warn in contrast to Floridi in a discussion of MacIntyre and Confucianism that the idealisation of communities in virtue ethics can advance authoritarian submission under hierarchies and "non-democratic social structures and practices" (46). Critical virtue ethics therefore needs to be based on a dialectic of the individual and society, agency and structures, subject and object in order to avoid individualism, provincialism, relativism, and authoritarianism. In the discussion of virtue cluster 4, Benkler and Nissenbaum (2006, 408) argue that MacIntyre "seems interested in the social contributions of virtues" because he gives attention to the virtue of creating a good life, but "does not develop in detail the relation between political and other communities and specific virtues".

Benkler and Nissenbaum argue based on MacIntyre (2007, 213) that humans face structural constraints, which in the context of commons-based peer production means that "incumbent firms" (Benkler and Nissenbaum 2006, 418) such as Microsoft tend to resist and oppose commons-based peer production. This argumentation implies that large corporations oppose the four clusters of virtues. But there is a dialectical complexity of the subsumption of aspects of society under capital: capital tries to subsume ever newer social systems in order to create new spheres of capital accumulation and to circumvent crisis tendencies and forestall resistance.

All of the virtuous practices discussed by Benkler and Nissenbaum are not automatically resisting the subsumption of the digital commons under capital. Someone involved in a peer production project can practice the virtues of autonomy, creative production, benevolence, charity, generosity, altruism, sociability, camaraderie, friendship, and co-operation by co-producing common resources, but can be subject to the exploitation of his or her labour for capital accumulation or can be the subject of such capital accumulation processes. Creativity, participation, sharing, openness, and co-operation have become new ideologies of digital capitalism: digital corporations such as Facebook, Google, for-profit open-access publishers, etc. practice the communism of capital: they advance the production of particularistic types of commons that are subsumed under the logic of capital. Facebook and Google accumulate capital through the free labour of users, who create, share, and participate in the production of data and content on platforms that are open for anyone to use as a gift. An ever-larger number of companies crowdsources product development, improvement and marketing to the free labour of the online crowd of brand consumers. Corporate open-access publishers accumulate capital by making content available as digital commons that are only released if the producers pay large sums of money as processing charges that not just finance end-production, but are also the source of corporate profits.

Benkler and Nissenbaum's four clusters of virtues of the commons are focused on individual and social virtues. They lack a third dimension: the dimension of collective political action (political virtues) that aims at creating a society of the commons and advancing struggles against the processes of commodification and bureaucratisation that subsume the commons under the logics of capital and domination. In order to advance a critical virtue ethics of the commons one needs to add the social struggle for a society of the commons as a fourth dimension of virtues. This means to politicise digital virtue ethics by adding the dimension of political virtues.

Such a critical dimension of virtues can however only be developed based on the concepts of essence and human nature. The most well-known version of MacIntyre's ethics is the phase, where he wrote *After Virtue*, in which he did not engage with the concept of human nature. However, in earlier and later phases of his philosophical development, he was more open to Karl Marx's critical theory that operates with the distinction between human essence and society's existence.

3.3 Human Essence

Burns (2011) distinguishes three subsequent stages in MacIntyre's works that he terms "Marx without Aristotle", "Aristotle without Marx", and "Aristotle and Marx". When MacIntyre published his first book *Marxism: An Interpretation* in 1953, he was "both a Christian and a Marxist" (MacIntyre 2009, 419). In 1968, when the book's second edition was released, he was neither a Christian nor a Marxist (ibid.). "During the years 1977 through 1984 MacIntyre transitioned to an Aristotelian worldview, returned to the Christian faith and turned from Aristotle to Thomas Aquinas" (Lutz 2015). In his major work *After Virtue*, MacIntyre wrote in 1981 that Marxism was "exhausted as a *political* tradition" and was politically too optimistic (MacIntyre 2007, 304). As part of his turn away from Marxism, MacIntyre gave up the concept of human nature (see MacIntyre 2002, 259, 261; 2007, 190). He later changed his position on that question and "became a Thomist after writing After Virtue" because "I became convinced that Aquinas was in some respects a better Aristotelian than Aristotle" (MacIntyre 2007, xi). In his late works, MacIntyre has returned to Marx and advances a "Thomistic Aristotelianism" that is "informed by Marx's insights" in order to construct "a contemporary politics and ethics" (MacIntyre 2016, xi).

Aristotle (1999, 1933) spoke of τὸ τί ἦν εἶναι (to ti ên einai) and τὸ τί ἐστι (to ti esti), which literally means "what it is in order to be" and "what it is". These phrases have either been translated as "essence" (e.g. in Aristotle 1933) or as "what it is for some-thing to be" (Aristotle 1999). The essence of an entity consists of "those characteristics that make it the kind of thing it is (or the very thing it is) and without which it could not exist or be what it is" (Meikle 1985, 177). Aristotle (1999, §1029b) defines essence as "what is *said* of" a thing "in its own right" (Aristotle 1999, §1029b [translated as "that which it is said to be per se" in Aristotle 1933, 1029b]). Essence means that something is a "primary thing" that is "not articulated by attributing one thing to another" (Aristotle 1999, §1030a; [translated as "do not involve the predication of one thing of another" in Aristotle 1933, §1029b]). The essence of a thing is "the substance which is peculiar to it and belongs to nothing else", whereas "the universal is common; for by universal we mean that which by nature appertains to several things" (Aristotle 1933, §1038b; [trans-lation in Aristotle 1999, §1030a:

> "For in the first place, the thinghood of each thing is what each is on its own, which does not belong to it by virtue of anything else, while the universal is a common property, since what is meant universally is what is of such a nature as to belong to more than one thing])".

Meikle (1985) distinguishes Aristotelian essentialism from the worldview of atomism. Ernst Bloch (1963, 1972) characterises Aristotle's concept of matter as a dynamic form of being-in-possibility *dynámei ón, δυνάμει ὄν*) and objective possibility. He opposes this concept of matter to mechanical materialism and traces it in the works of Avicenna, Averroes, Giordano Bruno, Spinoza, Schelling, Hegel, and Marx.

Justifying that the good life means a commons-based society requires reasonable as-sumptions about the nature of humans and society. MacIntyre argues in his early Marxist writings – especially the 1958/1959 essay *Notes from the Moral Wilderness* (MacIn-tyre 2009, 45–68) that for Marxists the "concept of human nature [...] has to be at the centre of any discussion of moral theory" (MacIntyre 2009, 63). For MacIntyre, morality has to do with human desires. Capitalism and class societies would distort desires so that there is a "rift between morality and desire" (61). MacIntyre here applies the He-gelian dialectical logic of essence and existence. Modern society would have created conditions where "human possibility can be realized in a quite new way" (56). But these potentials are artificially suppressed so that they do not benefit humanity in common, but rather predominantly the ruling class. "Each age reveals a development of human potentiality which is specific to that form of social life and which is specifically limited by the class structure of that society" (64). Each "new form of exploitation [...] brings new frustrations of human possibility" (125). "The paradox of bourgeois society is that it at one at the same time contains both the promise of greatly enlarged freedom and the denial of that freedom" (126). Human nature "is violated by exploitation and its accompanying evils" (66). Only class struggle is able to realise "a common shared humanity" (64) and "the deeper desire to share what is common in humanity" and "to rediscover common desire" (65). MacIntyre argues that class society means alienation from common human desires for solidarity. For him, human essence has to do with the collective human desire for a good life for all.

As part of his return to Marx during his late period, MacIntyre (2016, Chapter 2) argues in his book *Ethics in the Conflicts of Modernity* that Marx employs key Aristotelian concepts: "essence, potentiality, goal-directedness" (MacIntyre 2016, 94).

> "For Marx, as for Aristotle, human agents can be understood only as goal-directed, and we can distinguish between those goals the pursuit of which will develop their human potentiality and those goals the pursuit of which will frustrate their development".
>
> (94)

What is the role of language and communication in human nature? MacIntyre (2007, chapter 15) argues in *After Virtue* that the unity of human life can only be obtained through conversations that create social relations. The human being is a "story-telling animal" (250; see also the discussion in Williams 2009). Conversations construct dramatic narratives that make human life unpredictable. At the same time, human life is guided by a telos – "a variety of ends or goals" (250). Because of human life's communicative character, virtue is for MacIntyre not purely internal to practices, but supports the "quest for the good" and "increasing knowledge of the good" (254). MacIntyre sees the good as the search "for the good life for man" (254). This definition is unsatisfactory because the goal of what someone or a group understands as a good life can for example include the enslavement of others or genocide. Virtue ethics should not abstract from society as totality and not simply focus on particular communities. MacIntyre argues that the good life varies historically and from group to group (255). Therefore, virtues oriented on the social are for MacIntyre limited to particular local communities.

> "[W]e all approach our own circumstances as bearers of a particular social identity. [...] I belong to this clan, that tribe, this nation. Hence what is good for me hast to be the good for one who inhabits these roles".
>
> (255)

The stress of "moral particularity" disregards what humans have in common and faces the danger of turning into moral relativism and moral particularism. Avoiding moral relativism requires the concept of human nature/essence.

MacIntyre (2007, 265) in *After Virtue* speaks of the "narrative understanding of the unity of human life". This is just another formulation for saying that humans are communicative, social beings. MacIntyre does not draw the conclusion that communication and community are part of the human essence because in *After Virtue* he rejects the concept of human nature. His account thereby is incompletely Aristotelian.

MacIntyre's approach can be turned into a full Aristotelianism by adding the notion of human essence: humans are in essence producing, communicative, social beings. Aristotle describes humans as ζῷον λόγος ἔχων (zōon logon echon). Hannah Arendt (1958, 27) writes that it is inadequate to translate this category as a rational animal. In Greek, λόγος (logos) means both rationality/reason and speech/utterance. This double meaning precisely describes human essence: the human is a rational/teleological, communicative being. Language and work as forms of production are means for reaching goals. In *Politics*, Aristotle writes: λόγον δὲ μόνον ἄνθρωπος ἔχει τῶν ζῴων – "man alone among the animals has speech" (Aristotle 2013, §1253a).

MacIntyre (2016, 26) argues that "the power of language use" distinguishes humans from animals. Language would have four crucial features: it enables reflection and justifications, enhances the communication of intentions and responses, makes envisioning alternative futures possible, and allows narrating stories (26–27). Language enables humans to pose ethical questions about what is good (225).

Language has syntactic (form, rules), semantic (meaning, content), and pragmatic (effect, purposeful use in social contexts) aspects. MacIntyre (1999, 50–51) in his book *Dependent Rational Animals* discusses the example of bottlenose dolphins. These dolphins are highly developed animals that perform perceptive learning and communicate intentionally with each other. They form social bonds via a range of whistling sounds. MacIntyre uses the example of the dolphin to show that there are common biological features of highly developed animals and humans. Certain animals (dolphins, dogs, gorillas, chimpanzees, elephants, etc.) make use of prelinguistic means for achieving goals. Humans in contrast to animals are able to use language in complex manners in order to "express the judgement about which the agent is reflecting" (54), which allows them to reflect on and realise alternative actions (96). MacIntyre calls this capacity practical rationality (54). Humans are able to put language to reflective use (58). Through communicative social

Human Essence

relations, they learn to evaluate, modify, and reject their judgements (83) and to reflectively organise desires and the quest for wants and needs in order to achieve a variety of goods (96). Humans are therefore also moral beings that live through communication: "As a practical reasoner, I have to engage in conversation with others, conversation about what it would be best for me or them or us to do here and now, or next week, or next year" (110–111). For achieving the common good, humans have to not just communicate, but also need to co-operate (114).

So, humans strive to achieve individual and common goods by reflective and anticipatory judgement, learning through communicating judgements, practically enacting and modifying their judgements in everyday life, and working together with others. MacIntyre outlines common features of humans, but avoids speaking of "human essence". MacIntyre outlines a logic of essence but does not call it by this name. There are parallels of MacIntyre to Hegel, Marx, and Marcuse's concepts of essence (*Wesen* in German), which is why we should have a closer look at these thinker's works.

The German word *Wesen* has two meanings: it means (a) a creature or being; and (b) in a philosophical sense the particular features of a phenomenon that make it different from other phenomena and constitute the grounding and inner characteristics of something without which it could not exist (Duden 2019). Therefore, the two key English translations of *Wesen* are (a) create/being and (b) essence (Langenscheidt 2019). Especially, Hegel advanced the philosophical notion of *Wesen* by creating a dialectical logic of essence. For Hegel (1830/1991, §§115–130), the essence is the ground of existence. He speaks of a dialectic of essence (*Wesen*) and appearance (*Erscheinung*), which means that the essence of something that exists is not always immediately apparent. "The immediate being of things is [...] a sort of rind or curtain behind which the essence is concealed" (Hegel 1830/1991, Addition to §112). Actuality (*Wirklichkeit*) is for Hegel the dialectical sublation and unity of the contradiction between essence and appearance. "Actuality is the unity, become immediate, of essence and existence" (Hegel 1830/1991, §142). Actuality is a reasonable being, being not as it is immediately, but the way it can and should be so that it accords to the potentials inherent in its essence.

Herbert Marcuse (1936/2009) argues that Marx was heavily influenced by Hegel's dialectical logic of essence. Marx

> "works with two different sets of concepts, [...] One set describes the economic process in its immediate appearance. [...] the second group of concepts, which has been derived from the totality of the social dynamic, is intended to

grasp the essence and the true content of the manifestations which the first group describes as they appear. The dialectical concepts transcend given social reality in the direction of another historical structure which is present as a tendency in the given reality".

(Marcuse 1936/2009, 62–63)

Marx dialectically relates the two meanings of *Wesen* in his critical analysis of society and capitalism. On the one hand, Marx speaks of humans as species-being (*Gattungswesen*), by which Marx means that humans are producing, co-operative, social, and societal beings. "[P]roduction is his active species-being" (Marx 1844d, 277). The human species-being's "*species-powers*" lie in "the co-operative action of all [hu]mankind" (Marx 1844d, 333). The human being is "a societal animal" ("gesellschaftliches Tier", translation from German, Marx 1857/1858, 84).

For Marx, societality is the essence of human beings. He argues that in capitalism and class society, human societality is crippled and incomplete. He expresses this circumstance with the notion of alienation/estrangement (*Entfremdung*). Marx argues that human beings' "social activity" has the potential to create "the *true community*" of humanity, [...] but as long as man does not recognise himself as man, and therefore has not organised the world in a human way, this *community* appears in the form of *estrangement,* because its *subject,* man, is a being estranged from himself" (Marx 1844b, 217). In Marx's original German manuscript, the phrase "true community" is "das *wahre Gemeinwesen* der Menschen" (Marx 1844a, 451). The passage can best be translated as "the true commonwealth of humanity". But *Gemeinwesen* also has the word *Wesen* in it. The etymology of the word goes back to the combination of the two German words *gemein* (common) and *Wesen* (essence).[1] For Marx, the use of the term *Gemeinwesen* refers to the common as the communist existence of humans and the commons as the essence of humanity and society. So what Marx hints at in this and other passages is that class society alienates humans from the potential of the common control of society, a commons-based society.

In a passage in the *Economic and Philosophic Manuscripts*, Marx directly refers to communism as the actuality of society that realises the essence of the human being and society. Marx (1844d, 296) writes that communism is "the *positive* transcendence of *private property* as *human self*-estrangement" and therefore "the real *appropriation* of the *human* essence ['wirkliche Aneignung des menschlichen Wesens' in the German original – Marx 1844e, 536] by and for man"; communism therefore is "the complete return of man to himself as a *social* (i.e., human) being", "equals humanism" and is "the true resolution of the strife between existence and essence, between objectification and self-confirmation, between freedom and necessity, between the individual and the species". In this passage, Marx dialectically links the notion of *Wesen* as being and essence. Marx sees the commons as human essence, from which human beings are alienated in class societies. Communism sublates this antagonism between human essence and existence.

MacIntyre (2016) does not ask what the common aspects of the four dimensions of linguistic capacities that he identifies are, although he hints at the fact that language and communication "make possible kinds of cooperation and forms of association that are distinctively human" (26) and "enable us to associate cooperatively with others in ways not open to nonhuman animals" (29): humans produce social relations and produce in social relations. And they do so purposefully, namely, in order to try to advance human flourishing. Class relations and dominations however influence the perception of which human individuals and groups should flourish and which ones should be harmed as well as corresponding practices.

For Aristotle, human life in society has a telos, it is goal-oriented. Influenced by Aristotle and Marx, Georg Lukács (1978, 5) argues that the human being is a teleological being because the human as "conscious creator" produces with a purpose, orientation, and goal. Lukács sees teleological positing as the essence of the human being and society. Human beings are in essence teleological: they work and communicate in order to reach defined goals. Aristotle (2002, §1139b) argues that "one who makes something always makes it for the sake of something". Lukács combination of Aristotle and Marx shows that without production and communication there can be no human existence. Communication is the production of social relations and sociality. Work and communication are two aspects of production: humans produce meanings and sociality through communication. They produce use-values that satisfy human wants, needs, and desires through work. Work has a communicative character and communication work character. Work and communication are two dialectically encroaching dimensions of the practical, anticipatory, and reflective rationality of human beings.

Production in society is social and therefore communicative. Communication is productive: it creates shared understandings and sociality. Communication and work are common features of all humans and all societies. In order to attain desires and satisfy needs, we need to engage in production and communication so that immediate desires are suppressed and rationally transformed into work processes that allow attaining identified ends having to do with ways of how we can achieve wants, needs, and desires.

Production and communication require "intellect fused with desire" and "desire fused with thinking" (Aristotle 2002, §1139b). Humans can "stand back from our desires and other motives" (MacIntyre 2016, 44) in order to reflect and rationally act so that we achieve larger goals or what MacIntyre (2016, 53) characterises as the "final end" and the "ultimate human end", the human flourishing and good life that Aristotle characterises as eudaimonia. This means that humans have an essential quest for human flourishing and being able to lead a good life. If humans in essence desire the good life and are rational, communicating, producing, social, and societal animals who cannot achieve goals alone but only together with others, then the question arises of how humans can advance the common good in society.

3.4 The Common Good

In ethics, the concept of the common good is the one that stands closest to the notion of the commons. Therefore, if we want to ethically justify commons and communication commons, then the notion of the common good is the right philosophical starting point.

John Rawls defines the common good as "certain general conditions that are in an appropriate sense equally to everyone's advantage" (Rawls 1999, 217). Why should one care about the common good?

> "Many philosophers believe that there is something morally defective about a private society. [...] [The] members of a political community have a *relational obligation* to care about their common affairs, so the fact that they are exclusively concerned with their private lives is itself a moral defect in the community. [...]. A conception of the common good provides us with an account of what is missing from the practical reasoning of citizens in a private society, and it connects this with a wider view about the relational obligations that require citizens to reason in these ways. [...] Members of a political community stand in a social relationship, and this relationship also requires them to think and act in ways that embody a certain form of mutual concern. The common good defines this form of concern".
>
> (Hussain 2018)

The "common good has played an important role in Western political thought since its beginnings in ancient Greece" (5). Philosophers who wrote about the common good include Plato, Aristotle, Thomas Aquinas, John Locke, J.J. Rousseau, Adam Smith, G.W.F. Hegel, John Rawls, and Michael Walzer (Hussain 2018, Jade 2017). Aristotle is "a foundational thinker" of the common good (Jade 2017, 2). Given this foundational character and the importance of Aristotle for thinking about the common good, it makes sense to take a closer look at the Aristotelian concept of the common good.

Hussain (2018) distinguishes between joint activity- and private individuality conceptions of the common good. The first argues that members of a community have common interests that arise from their membership of and joint activities in the community. Representatives include Plato, Aristotle, Charles Taylor, or Michael Sandel. Private individuality theorists of the common good argue that members of a political community have an interest to guarantee certain common goods (such as liberal freedoms, democracy, the rule of law, internal, and external defence) in order to lead lives as private individuals. Representatives are for example Jean-Jacques Rousseau, Adam Smith, Georg W.F. Hegel, John Rawls, and Michael Walzer.

The first approach is more social, communal, and collectivist, the second is liberal-individualist. MacIntyre (1997) further differentiates these two approaches to the common good into communitarianism, liberalism, and Aristotelianism. He sees Aristotelianism not as a form of communitarianism, but a particular form of ethics that is opposed to both communitarianism and liberalism.

MacIntyre (1997) argues that the common good can be defined either as the end of community members' "shared activities" (239), or as the sum of individual goods in an association (239–240), or as activities in a polis, where individual and common goods are inseparable (241). For MacIntyre, both the first (communitarian) and the second (liberal) concept fail because in the modern state, liberal individualist and minimalist concepts and elements of the common good come into conflict with the social good that is based on a communitarian concept of the common good. MacIntyre recommends "small-scale local community politics" (248) that enables "a community of enquiry and learning" (251). MacIntyre does not take into account that the common good understood as a "community in which each individual's achievement of her or his own good is inseparable both from achieving the shared goods of practices from contributing to the common good of the community as a whole" (240–241) relates not just to local community practices, but also to practices that concern humanity as a whole, i.e. practices that either affect all humans or that are common to all humans. The common good certainly needs to take into account politics (the polis as a political community) and culture (common learning). Another important dimension is the common in the economy (common production, common ownership, and common access).

MacIntyre (2016) gives a range of examples for work towards the common good at the lo-cal level. "The common good of those at work together are achieved in producing goods and services that contribute to the life of the community and in becoming excellent at producing them" (MacIntyre 2016, 170). Family members

> "pursue their goods as family members by enabling the other through their affection and understanding to achieve her or his goods. Parents pursue their goods as family members by fostering the development of the powers and vir-tues of their children, so that those children may emerge from adolescence as independent rational agents".
>
> (169)

In schools, teachers

> "achieve their own good qua teachers and contribute to that common good by making the good of their students their overriding good, while their students contribute to the shared education of their class by their class participation, so achieving their own good".

Humans are rational, ethical, communicating, producing, social, and societal beings who behave purposefully in order to try to achieve a good life. Aristotle saw that there is an inherent connection of the commons, communication, and community. Communi-cation creates common meanings and definitions within a community. In modern class societies, the common good is subordinated under the logics of capital and bureaucracy. As a result, particularistic interests rule so that inequalities and asymmetric distribu-tions of power are a reality. The good life is not an actual common feature of all humans living in capitalism and class societies, where it is only a feature that some enjoy at the expense of others: some are forced by economic, political, or ideological structures to lead damaged, alienated lives. The commons are part of the human essence because the common features of all humans constitute the human essence. The desire for a good life is a common feature of all humans. But given that humans are social beings living in society, the good life cannot be achieved individually, but only collectively, socially, and politically. I can only lead a good life if all are enabled to lead good lives. A good society is a society that corresponds to human essence, i.e. a society of the commons, in which humans control the economic, the political and the cultural system, goods, and structures that together form the society in common so that everyone is empowered to lead a good life. A good society is a society of the commons. Alienation in contrast

The Common Good

means that humans are not in control of economic, political, and cultural structures that shape life in society.

Aristotle distinguishes between distributive, corrective, reciprocal, and universal justice (that advances the common good) (McCarthy 1990, Chapter 2). Justice and injustice are for Aristotle matters of proportionality and disproportionality. An "unjust person has more, while the one to whom injustice is done has less of something good" (Aristotle 2002, §1131b). Injustice means that someone has "an excess for oneself of what is simply beneficial and a deficiency of what is harmful" (§1134a). So Aristotle argues that injustice means that a certain individual, group or class has a kind of surplus control of a good over others. He anticipates Marx concept of surplus-value, but defines the excess he speaks of as a more general form of injustice that arises from the asymmetric distribution of power that benefits the few at the expense of the many. Marx (1867, 73–74) stresses that Aristotle was the first thinker who analysed the commodity's value-form. According to Marx (1867, 74), Aristotle was a genius because he saw that the value of commodities is a relation of equality. But given that Aristotle did not live in a capitalist society, he was not able to see that this equality is the effect of the objectification of specific quanta of labour in the commodity.

Aristotle (2002) not just opposes injustice to justice, but also to friendship and love, which are social relations where humans benefit and do good things for others without instrumental interests. The common arises from friendship and community: in

> "every sort of community there seems to be something just, and also friendship. [...] To whatever extent that they share something in common, to that extent is there a friendship, since that too is the extent to which there is something just. And the proverb 'the tings of friends are common' is right, since friendship consist in community. All things are common to brothers and comrades".
>
> (§1159b)

The political community aims at an advantage "that extends to all of life" (§1160a). Aristotle (2013, §1279a) terms a community where "the multitude governs with a view to the common advantage" polity.

Marx is an Aristotelian in respect to the good life as the realisation of human potentiality. Aristotle (1999, §1048a) sees potentiality as being "capable of something" and being "capable of causing motion". Potency is also the source of dialectic because whatever is potential "is itself capable of opposite effects" (Aristotle 1999, §1051a). In communism, the full potentials of human beings and society are actualised. Marx defines communism as a just society based on the commons, friendships, and love in Aristotle's understanding. One of Marx's achievements was that he uncovered how class societies in general and capitalism in particular structurally institutionalise the exploitation of labour and the injustices Aristotle spoke of. In essence, humans are co-operative, social, societal beings, who strive for solidarity and a good life. A particular societal condition enables or hinders the realisation of society's and human potentials. Such potentials develop historically. If class structures and domination make society's essence and existence diverge, then humans ought to organise collectively and struggle against alienation in order to realise a good society. Marx calls the good society "communism". His approach is teleological because humans have the potential to struggle for a good society. If they do so, then their social action becomes teleological – it becomes praxis, political action for the good society. Marx's Aristotelianism for example becomes evident when he argues that "our species-being [...] is not actualised as *energeia* in the context of private property" (Groff 2015, 316). In his political-economic works, Marx analyses the alienation of the human being from the essence of its species-being as abstract labour's creation of surplus-value, i.e. the exploitation of labour in class relations. The "proposition that man's species-nature is estranged from him means that one man is estranged from the other, as each of them is from man's essential nature" (Marx 1844d, 277). The German term for species-being is *Gattungswesen* (Marx 1844e, 518), a combination of species (*Gattung*) and the dual meaning of *Wesen* as essence/being. So, by species-being, Marx means human essence or what all human beings have in common. Capitalism and class constitute the alienation of the human from their social essence.

The commons are goods that all humans require in order to live a good life. The good life of the individual is only possible in a good society that enables the good life for all. Achieving a good society that benefits all requires collective organisation of the common good. It requires inclusive, co-operative communication. If structures of domination damage certain groups so that they are compelled to live alienated lives, then the common good is not realised. The good society is then only a good society for some – it is a class society. Establishing a good life therefore requires struggles and practices that are guided by "the *categorical imperative to overthrow all relations* in which man is a debased, enslaved, forsaken, despicable being" (Marx 1844c, 182). Without being able to live a good life in a society of the commons, humans are not fully developed humans – their existence does not correspond to their essence because they are denied those common goods that humans and society require to flourish and thereby realise their potentials.

The Common Good

MacIntyre argues in his later works that capitalism damages human life in multiple respects. Financial and educational inequalities would result in political inequalities (MacIntyre 2016, 127). Capitalism alienates human life in manifold respects. As a consequence, the institutions of the market, the state, and Morality shape human wants so that what individuals "want is what the dominant social institutions have influenced them to want" (MacIntyre 2016, 167).

MacIntyre (1999, 156) argues that we can only achieve the common good if our "social relationships of giving and receiving" are governed by "social and political forms" that advance the common good. He argues that three conditions must be fulfilled: (1) Institutionalised forms of deliberation are needed so that "shared rational deliberation" allows taking common decisions; (2) justice needs to be enabled so that each is working and giving "according to her or his ability" and each receives "so far as is possible, according to her or his needs" (130), which would have to especially take into account "those who are most dependent and in most need of receiving – children, the old, the disabled" (130); (3) everyone should "have a voice" in the community (130). Taken together, such a society of the commons advances the political commons (participatory democracy), the economic commons (wealth and self-fulfilment for all), and the cultural commons (voice and recognition of all). It requires to overcome political alienation (domination), economic alienation (exploitation), and cultural alienation (ideology).

MacIntyre (2016) argues that achieving the common good in communities such as families, workplaces, schools, or a political system depends on the availability of resources such as money, power, wealth, education/skills, and public goods provided by the government (the education system, law, order, health care, transport, communications, etc.). Capitalism is not just an economic system, but a type of society, in which the accumulation of money, power, and reputation results in structures that benefit certain groups at the expense of others:

> "The exploitative structures of both free market and state capitalism make it often difficult and sometimes impossible to achieve the goods of the workplace through excellent work. The political structures of modern states that exclude most citizens from participation in extended and informed deliberation on issues of crucial importance to their lives make it often difficult and sometimes impossible to achieve the goods of local community. The influence of Morality in normative and evaluative thinking makes it often difficult and sometimes impossible for the claims of the virtues to be understood, let alone acknowledge in our common lives".
>
> (237–238)

Given that humans are story-telling animals, they can learn "to *live* against the cultural grain" and "to act as economic, political, and moral antagonists of the dominant order" from

> "the stories of those who in various very different modern social contexts have discovered what hat to be done, if essential human goods were to be achieved, and what the virtues therefore required of them, so making themselves into critics and antagonists of the established order".
>
> (MacIntyre 2016, 238)

Resistance and social struggle require the communication of stories about how domination damages human life and how resistance is possible.

3.5 The Digital Commons and Morality

The outcome of the discussion in this chapter is that only a society that fosters the common good for all is a society appropriate to human beings, a morally and politically just society. Praxis is the ethico-political quest and struggles for a society of the commons. If ethics is consciousness and action oriented on the social struggle for a free society, then ethics aims at the point at which the "struggle for liberation changes dialectically into freedom" (Lukács 1971, 42). In such struggles, individuals overcome the isolation, separation, alienation, and partiality of their existence imposed by domination. They organise as political collectives that strive for freedom. "Praxis becomes the form of action appropriate to the isolated individual, it becomes his ethics" (Lukács 1971, 19).

Computers and computer networks enable new ways of organising information, communication, and co-operation. Given that computing has become a central resource in modern society, the use of computers for organising cognition, communication, and co-operation has become part of human needs. Humans have certain cognitive needs (such as being loved and recognised), communicative needs (such as friendships and community), and co-operative needs (such as working together with others in order to achieve common goals) in all types of society. In a digital and information society, computers are vital means for realising such needs. But given that computers are always used in societal contexts, computer use as such does not necessarily foster the good life, but can also contribute to damaging human lives. When it was shown in the Cambridge Analytica scandal that Facebook and other social media have been used for targeting users with fake news in order to try to manipulate elections, it became evident how a specific organisation of online platforms – namely the combination of digital capitalism, authoritarian politics, and neoliberalism – threatens the common political good of democracy. Advancing the good life for all with the help of computers requires a particular organisation and design of computing resources and society.

Combining insights from Aristotle, Karl Marx, and Alasdair MacIntyre helps us to argue why advancing the digital commons is morally important. Humans are in essence moral, rationally producing, communicating, social, and societal beings, who can only achieve their goals in relation to other humans. We can only achieve individual goals together with others. If achieving individual goals damages the life of others by exploiting or dominating them, then the common good is damaged because society will entail groups of humans who are compelled to lead damaged lives. A good society enables the good life for all. It is a commons-based society.

In what contexts do computers help to advance the good life or damaged lives? This question can be answered in respect to the use of computers at the level of society, i.e. applications of computing resources that affect all members of society. The important criterion for assessing computing ethically is the question if, how, and to which degree computers are used for advancing a good economic, political, and cultural life for all or are used for damaging economic, political, and cultural lives.

The economy has to do with questions of production and ownership. As economic beings, humans strive for a life that guarantees the satisfaction of their needs and allows self-fulfilling work. If computer resources that are vital for the life of all are commodities, then the lives of humans are negatively impeded in two ways: (1) Many commodities are produced by human labour that is exploited in class relations so that ownership is transferred from the immediate producers to private property holders, who obtain benefits at the expense of workers; (2) Goods and services that are exchanged as commodities will inevitably exclude those who cannot afford them from access. Given that exchange is always an unequal exchange, commodity-producing societies advance distributive injustice. Wikipedia as digital commons is preferable to a for-profit online encyclopaedia that sells access to articles in two respects: (1) A for-profit encyclopaedia will tend to rely on the exploitation of the human labour of digital workers who write or contribute to encyclopaedia articles in order to create profits; (2) Charging access to encyclopaedic resources will exclude humans from access. The exploitation of digital labour and denial of access to key digital resources damage the lives of humans economically. The digital commons are in contrast to digital capital inclusive and not class based.

An increasing number of for-profit companies rely on pseudo-commons in order to produce digital capital. They provide gratis access to certain digital resources, crowdsource human labour so that it is performed online and unpaid, and accumulate capital in so-phisticated manners that make humans not always experience their exploitation in an immediate manner. The digital commons thereby become subsumed under digital capi-talism. Advancing the digital commons therefore not just needs to entail advancing and supporting projects that foster the digital commons, but also to struggle against digital capital and digital capital that disguises itself as digital commons. Advancing the digital commons in a capitalist society as transcendent projects that prefigure an alternative mode of society faces the problem that humans in capitalism depend on wages in order to survive. The digital commons challenge digital capital and along with it to a certain degree also forms of wage-labour subsumed under digital capital. Advancing the digital commons as a class struggle project therefore requires mechanisms that guarantee that humans obtain income in order to survive and at the same time are empowered to act as digital communards. Mechanisms that tackle this problem include for example public/commons-partnerships, collective funding mechanisms, participatory budgeting, the tax-ation of corporations in order to fund alternative media projects, or basic income.

Politics is a system that organises the process of taking collective decisions that are binding for all members of society. As political beings (citizens), humans strive to in-fluence political decisions based on their interests. For doing so, legal systems that or-ganise rights, responsibilities, and freedom are necessary. Political life is damaged if (a) particular individuals or groups centralise the decision-making process so that other citizens or groups of citizens cannot participate in it or have less influence; or (b) rights that enable effective political voice in the public sphere or participation in collective decision-making are limited.

In the digital realm, authoritarian regimes use state power in order to censor the ex-pression of political voices online. They censor the publication of political information online, or monitor citizens and political opponents, or sanction those who express al-ternative opinions and dissent with the help of fines, prison sentences, or terror. As a consequence, political decision-making and political expression are centralised and the common political rights and interest of citizens to participate in decision-making are un-dermined. Economic power can undermine the common political good: if rich individuals, groups, or companies can use money in order to purchase political voice in the online public sphere (for example via online advertising or the ownership of popular news plat-forms) or to influence policies that govern the way digital resources are organised, then corporate power undermines the capacity of humans to voice their political opinions and to influence political decisions. Capitalist power threatens the common political good.

The Digital Commons and Morality

Using computing resources for fostering the political good requires the support of projects that aim at using digital resources for advancing participatory democracy. Participatory democracy aims at forms of empowerment that include all to a meaningful degree in political decision-making and fosters a public sphere, where inclusive, sustained political debate is possible and is not limited by hierarchies that are based on the unequal control of wealth, power, skills, and reputation.

Culture is the system in society that enables humans to make meaning of society and to define their identities, which requires voice, visibility, mutual understanding, and recognition. As cultural beings, humans strive for recognition. In the realm of online culture, Twitter is an example of a platform that humans use in order to make meaning of contemporary societies. But voice, visibility, reputation, and recognition are unequally distributed on Twitter: Whereas the average Twitter user has 707 followers,[2] the singers Katy Perry and Justin Bieber were in July 2018 with 110 million followers the two users with the largest amount of followers. Celebrities and corporations have a much higher online visibility, attention, voice, reputation, and recognition and therefore also more definition power on the Internet than everyday users. The asymmetric distribution of voice and visibility damages human lives because it denies humans the capacities for recognition and influencing collective meaning-making processes in society. The result is cultural hierarchies, in which influencers have much more power to be heard and shape worldviews than everyday people.

Today, also online bots generate reputation by posting and re-posting stories, liking social media postings, etc. Automated programmes thereby generate fake attention and operate fake social media profiles. Machines do not have morals and therefore the problem of fake online attention generated by bots is that they try to make humans believe that a machine response and behaviour is a form of human recognition. Imagine there is an intelligent bot that through sustained communication with individuals creates feelings of love online. When the user finds out that the online interlocutor is an artificial intelligence-based machine who "cheats" on him/her with hundreds of other users, s/he will be disappointed and feel tricked because machines can only simulate, but not experience emotionality and morality. S/he will feel cheated because s/he invested time into building an emotional relationship and has to find out that this work of love was not reciprocated by a human being, who spent time on affective expressions, but by a machine. The movie *Her* is exactly about the issue of "love" between humans and an artificially intelligent operating system named Samantha (to whom Scarlett Johannson gives a voice in the movie). In the age of bots and fake attention, it becomes difficult to discern what is communicated by humans and what by machines. Algorithms can manipulate perception.

Digital media that help fostering the cultural good help all humans to make their voices heard, to achieve common understandings, and to achieve recognition. Humans all strive for recognition, but have different worldviews, identities, and lifestyles. A common culture is not a unitary culture, but one that constructs the unity in the diversity of worldviews, ways of life, and identities that are needed for respect and understanding. A common culture avoids both the extremes of cultural imperialism (unity without diversity) and cultural relativism (diversity without unity).

Table 3.2 provides a summary overview of the dimensions of the digital commons.

The typology presented in Table 3.2 is structured along with the three realms of society (economy, politics, and culture), which allows distinguishing between three types of commons and three types of digital commons. The commons are the Aristotelian-Marxian vision of a good society. They form the essence of society, which means that the digital commons are part of digital society's essence. We discussed earlier in this chapter, that for Hegel and Marx the essence is often hidden behind false appearances and that actuality means the correspondence of essence and appearance. An Aristotelian–Hegelian–Marxian perspective on the digital commons therefore needs to distinguish between the essence of digital society and the false appearance and existence of digital society as digital class society and digital capitalism. Class society is the false condition of society in general. Digital class society is the false condition of digital society. An

TABLE 3.2 Three dimensions of the digital commons

	Commons in society	Digital commons	Lack of common control in society (alienation)	Lack of common control of digital society (digital alienation)
Economy	Economic commons: wealth and self-fulfillment for all	Economic digital commons: network access for everyone, community is in control of technology, digital resources as common goods	Private property	Digital commodities, digital resources as private property
Politics	Political commons: participation and democracy in decision-making	Political digital commons: common decision making/governance of ICTs	Dictatorship	Dictatorial governance and control of ICTs
Culture	Cultural commons: voice and recognition of all	Cultural digital commons: use of ICTs for fostering learning, recognition and community activities	Ideology	Digital ideology: Ideologies of and on the Internet

The Digital Commons and Morality

Aristotelian–Hegelian–Marxian ethics of the commons needs to not just have a vision of a good digital society, but is also a critique of digital capitalism and digital alienation. Table 3.2 therefore also features two columns that outline dimensions of alienation-in-general and digital alienation.

Aristotelian ethics differs from both deontology and consequentialism/utilitarianism. Deontology is an individualistic ethics focused on individual behaviour. Consequentialism focuses on the consequences of the action. In contrast, Aristotelian ethics is interested in ethical practices and learning processes. Marxian ethics that builds on Aristotle is interested in ethical practice as critical praxis. In digital capitalism, critical ethical praxis takes on three forms:

1) *Critical digital theory*: In the realm of academia, research, and intellectual life, Aristotelian–Hegelian–Marxian ethics challenges theories and approaches that fetishise and justify digital exploitation and digital domination and that fetishise instrumental reason, quantification, calculation as is for example the case in big data analytics (Fuchs 2017a). It generates systematic knowledge that wants to inform and empower class struggles against digital alienation and for a commons-based digital society, where all humans benefit from digital technologies.

2) *Critical digital education*: In the realm of education, Aristotelian–Hegelian–Marxian ethics challenges the pure focus on teaching quantitative skills (STEM: science, technology, engineering, mathematics) that aim at turning education into an instrument of digital capital's innovation and capital accumulation strategies and envisions all young learners as future entrepreneurs. It also challenges the metrification of education itself. In contrast, such an ethics fosters critical education that empowers humans' critical reason so that they are able to reflect on the complexities and causes of digital society's problems and understand the roots of digital capitalism's contradictions.

3) *Digital class struggles*: In the realm of politics, Aristotelian–Hegelian–Marxian ethics empowers humans to challenge digital alienation and to support and engage in class struggles that aim at establishing a fair, just and good digital society of the commons, where all benefit. Class struggles are struggles for the control of economic resources, working conditions, and economic decisions. In digital capitalism, class struggles have two digital aspects: (a) workers in the digital industry face precarity and exploitation and organise collectively in order to struggle against exploitation; (b) class struggles in general-use digital means of communication for their organisation and for public communication.

The next section will discuss three examples of how Aristotelian–Hegelian–Marxian ethics can be applied to questions of digital society.

3.6 Examples of the Application of Aristotelian–Hegelian–Marxian Ethics to Digital Media

In order to outline how Aristotelian–Hegelian–Marxian ethics operates, we will in this section focus on three examples, one each from the realms the digital economy, digital politics, and digital culture: digital labour, digital surveillance, and digital authoritarianism.

The first example focuses on labour in digital capitalism, specifically the digital labour that creates the profits of Google and Facebook. In 2018, Alphabet/Google accumulated profits of US$30.7 billion, which meant an annual increase by 142.7%. In 2018, Facebook's profits amounted to US$22.1 billion, an increase by 38.8% in comparison to 2017. In 2018, Alphabet/Google was the world's 23rd largest transnational corporation and Facebook the 77th largest.[3] With more than a billion monthly active users of Google search and YouTube[4] and 2.4 billion monthly active Facebook users,[5] the two companies are the world's largest advertising agencies.

According to Marx, capitalist corporations yield profit by selling commodities that contain unpaid surplus value that workers produce. What is the commodity that Facebook and Google sell? Access to the platforms is gratis, which means that the software that drives Google and Facebook are not commodities. The two companies sell targeted ads with the help of algorithmic auctions. Users' online activities of clicking and viewing are the value-generating labour that yields the profits of companies such as ad-financed digital corporations such as Google, YouTube, and Facebook. Users create data, meta-data, comments, searches, views, likes, information and communication flows, and social relations. The use of Google and Facebook is a form of digital labour that creates a data commodity and surplus value (Fuchs 2017b). What does digital alienation mean in the context of Google and Facebook? It means that the two companies exploit their users. All exploited labour has an unpaid part. In the case of wage labour, there is a paid and an unpaid part. In the case of Facebook and Google user labour, all of the labour time is unpaid labour time.

As a consequence, Facebook's CEO Mark Zuckerberg is the world's eight richest person, and Google's founders Larry Page and Sergey Brin are the tenth and fourteenth richest individuals.[6] Google and Facebook defy the logic of the digital commons because the two platforms are built on a class relationship between their owners on the one side and the users, who are digital workers, on the other hand. In digital capitalism, the borders between labour time and leisure time, labour and play, production and consumption, the private and the public sphere, the factory and society are blurred. How do digital platforms that benefit all look like? The digital commons in the realm of the digital economy means common ownership and control of digital platforms. Such platforms operate as not-for-profit businesses. Examples that could be established but do not yet exist are an alternative non-profit YouTube operated by a network of public service media and an alternative Facebook that is collectively owned by its users. Public service Internet platforms and platform co-operatives are two alternative models of the Internet (Fuchs 2018b; Scholz and Schneider 2017).

The Cambridge Analytica scandal forms the second example. It is a case from the realm of digital politics. According to reports by the *Guardian*, Cambridge Analytica paid money to the company Global Science Research (GSR) in order to carry out online personality tests on Facebook. As a consequence, personal data of around 90 million users[7] were collected, which included friendship data and likes (Cadwalladr 2018; Cadwalladr and Graham-Harrison 2018; Kang and Frenkel 2018; Rosenberg, Confessore and Cadwalladr 2018). The *New York Times* wrote that the Cambridge Analytica data breach "allowed the company to exploit the private social media activity of a huge swath of the American electorate, developing techniques that underpinned its work on President Trump's campaign in 2016" (Rosenberg, Confessore and Cadwalladr 2018). According to such news articles, the collected data was utilised in order to target political ads in various election campaigns, including Donald Trump's 2016 campaign to become the US president and campaigns propagating Britain's exit from the EU in the 2016 Brexit-referendum.

The Cambridge Analytica scandal is a story of how digital surveillance threatens democracy. Capitalist social media gather as much data as possible about users in order to sell targeted ads. They operate based on the principle "data means profit" and therefore have erected the world's largest surveillance machines that operate on the Internet. Right-wing authoritarian groups and movements see digital surveillance as a means for the political control of citizens and voters. Digital alienation means in this context that not users, but digital corporations govern what user data is collected, how long it is stored, and for what purposes it is used. In addition, political digital alienation also means that right-wing authoritarian movements try to win elections and referenda by making use of digital surveillance, targeted ads, and false information ("fake news"). From an ethical and legal perspective, data that stands in the context of politics, such as political and philosophical worldviews, are sensitive. Users, who sign up to a digital platform, do not expect that their online behaviour is subject to digital surveillance that is used for targeting them with right-wing propaganda and false information. How could an alternative look like? In the realm of digital governance, the digital commons mean that personal data collection and use is privacy-friendly and minimal and that users participate and have a say in the terms of use and privacy policies that govern data collection. Furthermore, such a critical perspective requires to overcome the dominance of corporate self-regulation of data collection practices and data protection laws that protect the interests of users, citizens, workers, consumers, and prosumers (Fuchs 2011).

Our third example stems from the realm of digital culture. It focuses on the spreading of nationalist and xenophobic ideology via social media. Let us have a look at the following tweet by Nigel Farage, who was the leader of the far-right UK Independence Party (UKIP) from 2006 to 2009 and from 2010 to 2016 and has since March 2019 led the Brexit Party.

Nigel Farage ✓
@Nigel_Farage

···

We must break free of the EU and take back control of our borders.

1:22 pm · 16 Jun 2016 · Twitter Web Client

618 Retweets   **50** Quote Tweets   **795** Likes

FIGURE 3.1   Tweet by Nigel Farage in the context of the Brexit-referendum.

Examples of the Application of Aristotelian–Hegelian–Marxian Ethics to Digital Media

Farage made this post seven days before the Brexit-referendum on his Twitter account that at the time when the posting was made had 321,000 followers.[8] It shows a BBC video interview, where Farage talks about a UKIP poster campaign that shows a mass of refugees and the slogans "BREAKING POINT The EU has failed us all. We must break free of the EU and take back control of our borders. Leave EU".

Digital alienation means in this context that Farage tries to present immigrants and refugees as the cause of social problems, which diverts attention from the class antagonism between capital and labour. The Tories and New Labour have since Thatcher's resignation as British Prime Minister in 1990 continued neoliberal, Thatcherite politics that has resulted in large inequalities. Farage claims that Britain's inequalities are caused not by the internal antagonisms of capitalism, but by external factors, namely, immigrants, refugees, and the EU. He constructs scapegoats in order to divert attention from the power structures that underlie social inequalities.

The UKIP poster depicts migrants and refugees as a mass and flood that reminds the audience of "a water-course/current/flood that has to be 'dammed'" (Reisigl and Wodak 2001, 59). To speak of migration and refugees as resulting in a "breaking point" combines two ideological strategies, the topos of danger, the topos of large numbers, and the topos of burdening (Reisigl and Wodak 2001, 77–79). The first topos claims that a certain political action (such as membership of the EU) results have "specific dangerous, threatening consequences" (Reisigl and Wodak 2001, 77), in this case, the break-up of British society and its social system. The second topos constructs a danger by communicating that a mass of migrants and refugees threatens Britain. The third topos communicates that mass immigration burdens Britain.

Right-wing authoritarian ideology is not built on facts, but on negative emotions, fears, unsubstantiated claims, and the stoking of prejudices and hatred. The falseness of UKIP's xenophobic ideology can relatively easily be deconstructed by looking at some facts: 26% of British doctors are foreign nationals; 9.7% of British doctors are EU citizens; 16% of the nurses and health visitors working for the NHS are non-British citizens; 6.8% of nurses and health visitors are EU nationals (Kentish 2018; UK Parliament 2008). Without migrant workers, the British health system would reach a "breaking point". In 2020, 28% of British citizens will have reached or will be older than the state pension age; in 2040 the share will have increased to 37% (Pensions Policy Institute 2019). In 2016, 83% of the first-time asylum seekers in the EU 28 countries were less than 35 years old and 51% were aged between 18 and 34 (data source: Eurostat). Ageing societies need migrants in order to prevent that their pension systems will reach "breaking points". Many more examples that show the vital importance of migrants for contemporary societies could be added.

Right-wing demagogues make use of social media in order to spread authoritarian ideology (Fuchs 2018a). In May 2019, Donald Trump had 60 million Twitter followers, 25 million Facebook followers, and almost 13 million Instagram followers; Marine Le Pen had 2.24 million Twitter followers, 1.5 million Facebook followers, and 110k Instagram followers; Nigel Farage had 1.3 million Twitter followers, 850k Facebook followers, and almost 70k Instagram followers. Farage's tweet shown in Figure 3.1 achieved more than 1,000 likes, 790 re-tweets, and 217 comments. The far-right uses social media for spreading authoritarian propaganda in the public sphere. The alternative to right-wing authoritarian ideology online is humanism online, i.e. the use of social media for challenging and unmasking ideology and advancing understanding and the recognition of humans and their importance for society.

The three examples show how digital alienation works. Critical digital praxis challenges digital alienation. Section 3.5 identified critical digital theory, critical digital education, and digital struggles as forms of critical digital praxis. Establishing alternatives to domination and exploitation in the age of digital capitalism requires a theory that unmasks ideology and opens up visions for a society of the digital commons, educational efforts that empower humans to critically understand and challenge digital alienation, and social struggles that establish alternative digital media and whose aim is a society that benefits all. These three levels of praxis are intertwined.

3.7 Conclusion

A contemporary Aristotelian–Hegelian–Marxist digital media ethics is based on the insight that fostering the digital commons is a way for advancing the common good and a good life for all humans. Virtuous commoners challenge, criticise, struggle against, and aim to abolish digital resources that advance exploitation (economic alienation), authoritarianism (political alienation), and ideology (cultural alienation). Their goal is to advance digital commons in a society of the commons so that economic, political, and cultural power are distributed in ways that benefit all. Communication requires community and the commons. A fully communicative society – a communication society that corresponds to human essence – is a community of commoners, a commons-based society, where the common good helps advancing individual goods and humans in pursuing individual goods help advancing the common good. The ethics of digital commons are not independent of the ethics of the commons. In order to advance the digital commons, we need to advance towards a society of the commons in general that forms the context for the digital commons. Struggles for advancing the digital commons have to be struggles for a commons-based society.

Brey's (2010) approach of disclosive computer ethics aims at disclosing how "morally opaque practices" (51) are present and hidden in and designed into computer technologies. It is based on the insight that technologies embed moral values and therefore have politics. Value-sensitive design (VSD) is a complementary approach that aims at designing computer technologies based on moral values such as privacy, freedom from bias, autonomy, human welfare, etc. (Friedman, Kahn and Borning 2008). VSD must take into account that technologies operate in a societal context and that their ethical redesign therefore needs to come along with political changes, i.e. the redesign of society. Otherwise, VSD turns into a techno-centric moralism that disregards society's contradictions. The point of commons-based design is to create commons-based technologies in commons-based social systems that are guided by the struggle for a commons-based society.

James H. Moor (1998/2000) asks what Aristotle would do if he were alive today and a computing professional. Aristotle would not just be an active contributor and supporter of Wikipedia and other digital commons projects, he would programme free software and create digital commons that are used in collective political struggles against digital capitalism and digital domination because he would view such structures as forms of oppression that limit human potentials, damage the good life and human flourishing. If Aristotle were alive today, he would be a digital commonist.

Notes

1   See Deutsches Wörterbuch von Jacob Grimm und Wilhelm Grimm. http://woerterbuchnetz.de (accessed on 9 September 2019): entry for "Gemeinwesen".
2   https://www.brandwatch.com/blog/44-twitter-stats/ (accessed on 1 July 2018).
3   Data source: Forbes Global 2000 list for the year 2018. https://www.forbes.com/global2000/list (accessed on 8 May 2019).
4   https://abc.xyz/investor/static/pdf/20180204_alphabet_10K.pdf?cache=11336e3 (accessed on 8 May 2019).
5   https://newsroom.fb.com/company-info/ (accessed on 8 May 2019).
6   Forbes' list of the world's billionaires 2018. https://www.forbes.com/billionaires/list/ (accessed on 8 May 2019).

7 Journalists first estimated that the data breach affected around 50 million Facebook users. A bit later, Facebook indicated that almost 90 million users' personal data may have been accessed and collected (Kang and Frenkel 2018).

8 Data source: https://web.archive.org/web/20160616163225/http://twitter.com/nigel_farage (accessed on 8 May 2019).

References

Adorno, Theodor W. 2001. *Metaphysics: Concept and Problems*. Cambridge: Polity.

Arendt, Hannah. 1958. *The Human Condition*. Chicago, IL: The University of Chicago Press.

Aristotle. 2013. *Aristotle's Politics*. Translated by Carnes Lord. Chicago, IL: The University of Chicago Press. Second edition.

Aristotle. 2002. *Nicomachean Ethics*. Translated by Joe Sachs. Indianapolis, IN: Hackett.

Aristotle. 1999. *Metaphysics*. Translated by Joe Sachs. Santa Fe, NM: Green Lion Press.

Aristotle. 1933. *The Metaphysics: Books I-IX*. Translated by Hugh Tredennick. London: William Heinemann Ltd.

Benkler, Yochai. 2006. *The Wealth of Networks: How Social Production Transforms Markets and Freedom*. New Haven, CT: Yale University Press.

Benkler, Yochai and Helen Nissenbaum. 2006. Commons-Based Peer Production and Virtue. *The Journal of Political Philosophy* 14 (4): 394–419.

Bloch, Ernst. 1972. *Das Materialismusproblem, seine Geschichte und Substanz*. Frankfurt am Main: Suhrkamp.

Bloch, Ernst. 1963. *Avicenna und die Aristotelische Linke*. Frankfurt am Main: Suhrkamp.

Brey, Philip. 2010. Values in Technology and Disclosive Computer Ethics. In *The Cambridge Handbook of Information and Computer Ethics*, ed. Luciano Floridi, 41–58. Cambridge: Cambridge University Press.

Burns, Tony. 2011. Revolutionary Aristotelianism? The Political Thought of Aristotle, Marx, and MacIntyre. In *Virtue and Politics: Alasdair MacIntyre's Revolutionary Aristotelianism*, ed. Paul Blackledge and Kelvin Knight, 35–53. Notre Dame, IN: University of Notre Dame Press.

Cadwalladr, Carole. 2018. "I Made Steve Bannon's Psychological Warfare Tool": Meet the Data War Whistleblower. *The Guardian Online*. March 18, 2018. https://www.theguardian.com/news/2018/mar/17/data-war-whistleblower-christopher-wylie-faceook-nix-bannon-trump

Cadwalladr, Carole and Emma Graham-Harrison. 2018. Revealed: 50 Million Facebook Profiles Harvested for Cambridge Analytica in Major Data Breach. *The Guardian Online*. March 17, 2018. https://www.theguardian.com/news/2018/mar/17/cambridge-analytica-facebook-influence-us-election

Callinicos, Alex. 2011. Two Cheers for Enlightenment Univerdalism. Or, Why It's Hard to Be an Aristotelian Revolutionary. In *Virtue and Politics: Alasdair MacIntyre's Revolutionary Aristotelianism*, ed. Paul Blackledge and Kelvin Knight, 54–78. Notre Dame, IN: University of Notre Dame Press.

Duden. 2019. *Duden German Online Dictionary: Wesen.* https://www.duden.de/rechtschreibung/ Wesen (accessed on 8 May 2019).

Ess, Charles and Hallvard Fossheim. 2013. Personal Data: Changing Selves, Changing Privacies. In *Digital Enlightenment Yearbook 2013*, ed. Mireille Hildebrandt, 40–55. Amsterdam: IOS Press.

Floridi, Luciano. 2013. *The Ethics of Information.* Oxford: Oxford University Press.

Friedman, Batya, Peter H. Kahn Jr., and Alan Borning. 2008. Value Sensitive Design and Information Systems. In *The Handbook of Information and Computer Ethics*, ed. Kenneth Einar Himma and Herman T. Tavani, 69–101. Hoboken, NJ: Wiley.

Fuchs, Christian. 2018a. *Digital Demagogue: Authoritarian Capitalism in the Age of Trump and Twitter.* London: Pluto Press.

Fuchs, Christian. 2018b. *The Online Advertising Tax as the Foundation of a Public Service Internet.* London: University of Westminster Press.

Fuchs, Christian. 2017a. From Digital Positivism and Administrative Big Data Analytics Towards Critical Digital and Social Media Research! *European Journal of Communication* 32 (1), 37–49.

Fuchs, Christian. 2017b. *Social Media: A Critical Introduction.* London: Sage. Second edition.

Fuchs, Christian. 2016. Information Ethics in the Age of Digital Labour and the Surveillance-Industrial Complex. In *Information Cultures in the Digital Age: A Festschrift in Honor of Rafael Capurro*, ed. Matthew Kelly and Jared Bielby, 173–190. Wiesbaden: Springer.

Fuchs, Christian. 2011. Towards an Alternative Concept of Privacy. *Journal of Information, Communication and Ethics in Society* 9 (4): 220–237.

Groff, Ruth. 2015. On the Ethical Contours of Thin Aristotelian Marxism. In *Constructing Marxist Ethics: Critique, Normativity, Praxis*, ed. Michael J. Thompson, 313–335. Leiden: Brill.

Hegel, Georg Wilhelm Friedrich. 1830/1991. *The Encyclopaedia Logic (With the Zusätze). Part I of the Encyclopaedia of Philosophical Sciences,* translated by Theodore F. Geraets, Wallis A. Suchting and Henry S. Harris. Indianapolis, IN: Hackett.

Hess, Charlotte and Elinor Ostrom. 2003. Ideas, Artefacts, and Facilities: Information as a Common-Pool Resource. *Law and Contemporary Problems* 66 (1/2): 111–145.

Hess, Charlotte and Elinor Ostrom. 2007. Introduction: An Overview of the Knowledge Commons. In *Understanding Knowledge as a Commons: From Theory to Practice*, ed. Charlotte Hess and Elinor Ostrom, 3–26. Cambridge, MA: The MIT Press.

Hussain, Waheed. 2018. The Common Good. In *Stanford Encyclopedia of Philosophy.* https://plato. stanford.edu/entries/common-good/

Jade, Maximilian. 2017. *The Concept of the Common Good.* Working Paper Series of the Political Settlements Research Programme. Edinburgh: University of Edinburgh. https://www.britac. ac.uk/sites/default/files/Jaede.pdf

Kang, Cecilia and Sheera Frenkel. 2018. Facebook Says Cambridge Analytica Harvested data of up to 87 Million Users. *The New York Times Online.* April 4, 2018. https://www.nytimes.com/ 2018/04/04/technology/mark-zuckerberg-testify-congress.html?rref=collection%2Fbyline %2Fmatthew-rosenberg

Kentish, Benjamin. 2018. How Reliant is the NHS on Foreign Doctors? *The Independent Online.* June 4, 2018. https://www.independent.co.uk/news/uk/politics/nhs-foreign-doctors-how-many-reliant-immigration-theresa-may-brexit-explained-visa-a8383306.html

Langenscheidt. 2019. *Langenscheidt German/English-dictionary: Wesen.* https://de.langenscheidt.com/deutsch-englisch/wesen#Wesen (accessed on 8 May 2019).

Lukács, Georg. 1978. *The Ontology of Social Being. 3: Labour.* London: Merlin.

Lukács, Georg. 1971. *History and Class Consciousness.* London: Merlin.

Lutz, Christopher Stephen. 2015. Alasdair Chalmers MacIntyre. In *Internet Encyclopedia of Philosophy.* http://www.iep.utm.edu/mac-over/

MacIntyre, Alasdair. 2016. *Ethics in the Conflicts of Modernity. An Essay on Desire, Practical Reasoning, and Narrative.* Cambridge: Cambridge University Press.

MacIntyre, Alasdair. 2011. Where We Were, Where We Are, Where We Need to Be. In *Virtue and Politics: Alasdair MacIntyre's Revolutionary Aristotelianism*, ed. Paul Blackledge and Kelvin Knight, 307–334. Notre Dame, IN: University of Notre Dame Press.

MacIntyre, Alasdair. 2009. *Alasdair MacIntyre's Engagement with Marxism: Selected Writings 1953–1974*, ed. Paul Blackledge and Neil Davidson. Chicago, IL: Haymarket.

MacIntyre, Alasdair. 2007. *After Virtue: A Study in Moral Theory.* Notre Dame, IN: University of Notre Dame Press. Third edition.

MacIntyre, Alasdair. 2002. *A Short History of Ethics.* Abingdon: Routledge.

MacIntyre, Alasdair. 1999. *Dependent Rational Animals. Why Human Beings Need the Virtues.* Chicago, IL: Open Court.

MacIntyre, Alasdair. 1997. Politics, Philosophy and the Common Good. In *The MacIntyre Reader*, ed. Kelvin Knight, 235–252. Cambridge: Polity Press.

MacIntyre, Alasdair. 1953. *Marxism: An Interpretation.* London: SCM Press.

Marcuse, Herbert. 1936/2009. The Concept of Essence. In *Negations: Essays in Critical Theory*, ed. Herbert Marcuse, 31–64. London: MayFly Press.

Marx, Karl. 1867. *Das Kapital. Kritik der politischen Ökonomie. Erster Band. MEW Band 23.* Berlin: Dietz.

Marx, Karl. 1857/1858. *Grundrisse.* London: Penguin.

Marx, Karl. 1844a. Auszüge aus James Mills Buch „Élémens d'economie politique". In *MEW Band 40*, 445–463. Berlin: Dietz.

Marx, Karl. 1844b. Comments on James Mill, Élémens d'economie politique. In *MECW Volume*, 211–228. London: Lawrence & Wishart.

Marx, Karl. 1844c. Contribution to the Critique of Hegel's Philosophy of Law. In *MECW Volume 3*, 175–187. London: Lawrence & Wishart.

Marx, Karl. 1844d. Economic and Philosophic Manuscripts of 1844. In *MECW Volume 3*, 229–346. London: Lawrence & Wishart.

Marx, Karl. 1844e. Ökonomisch-philosophische Manuskripte aus dem Jahre 1844. In *Marx-Engels-Werke (MEW) Band 40*, 465–588. Berlin: Dietz.

McCarthy, George E. 1990. *Marx and the Ancients: Classical Ethics, Social Justice, and Nineteenth-Century Political Economy.* Savage, MD: Rowman & Littlefield.

Meikle, Scott. 1985. *Essentialism in the Thought of Karl Marx.* London: Duckworth.

Moor, James H. 1998/2000. If Aristotle Were a Computing Professional. In *Cyberethics: Social & Moral Issues in the Computer Age*, ed. Robert M. Baird, Reagan Ramsower, and Stuart E. Rosenbaum, 34–40. Amherst, NY: Prometheus Books.

Ostrom, Elinor. 1990. *Governing the Commons. The Evolution of Institutions for Collective Action.* Cambridge: Cambridge University Press.

Pensions Policy Institute. 2019. Demographics. https://www.pensionspolicyinstitute.org.uk/re-search/pension-facts/table-1/ (accessed on 8 May 2019).

Rawls, John. 1999. *A Theory of Justice.* Cambridge, MA: Belknap Press. Revised edition.

Reisigl, Martin and Ruth Wodak. 2001. *Discourse and Discrimination. Rhetorics of Racism and Antisemitism.* London: Routledge.

Rosenberg, Matthew, Nicholas Confessore, and Carole Nicholas. 2018. How Trump Consultants Exploited the Facebook Data of Millions. *The New York Times Online.* March 17, 2018. https://www.nytimes.com/2018/03/17/us/politics/cambridge-analytica-trump-campaign.html

Scholz, Trebor and Nathan Schneider, eds. 2017. *Ours to Hack and to Own. The Rise of Platform Cooperativism. A New Vision for the Future of Work and a Fairer Internet.* New York: OR Books.

UK Parliament. 2018. NHS staff from overseas: statistics. *House of Commons Library.* https://researchbriefings.parliament.uk/ResearchBriefing/Summary/CBP-7783

Whyman, Tom. 2017. Adorno's Aristotle Critique and Ethical Naturalism. *European Journal of Philosophy* 25 (4): 1208–1227.

Williams, Bernard. 2009. Life as Narrative. *European Journal of Philosophy* 17 (2): 305–314.

Part II

Applications

Chapter Four

Information Ethics in the Age of Digital Labour and the Surveillance-Industrial Complex

4.1 Information Ethics: Capurro and Floridi

For Rafael Capurro (2003, 22), information ethics poses questions about the Enlightenment in the information age. It asks, "How can we ensure that the benefits of information technology are not only distributed equitably, but that they can also be used by the people to shape their own lives?" (41).

> "Information ethics as a *descriptive theory* explores the power structures influencing attitude towards information and traditions in different cultures and epochs. Information ethics as an *emancipatory theory* develops criticisms of moral attitudes and traditions in the information field at an individual and collective level".
>
> (198)

It explores and evaluates "the development of moral values in the information field, the creation of new power structures in the information field, information myths, hidden contradictions and intentionalities in information theories and practices, the development of ethical conflicts in the information field" (198).

Solving these tasks would require that information ethics both thinks about institutional design and cares about the self's needs, such as friendship, respect, social relations, silence, laughter, etc. (43). Capurro's approach stresses the need for information ethics to pay attention to information technology's ambiguities in society (39–41), such as the information gap, technological colonisation, cultural alienation, or oligarchic information control (42). It also inquires into the tensions between freedom of communication/privacy, free online culture/copyright, the information rich and the information poor, information markets/digital democracy, the global and the local online community, oneness

DOI: 10.4324/9781003279488-6

and unity/diversity and plurality online (144). It questions "structures of power and oppression" (ibid.).

Although he will not agree with my analysis because, based on his view of Heidegger's position, he tends to see Hegel and Marx as representatives of a deterministic and totalitarian metaphysics that conceives of history as necessary progress (24–28), Capurro advances a concept of information ethics that in its stress on ambiguities of the information age is not unrelated to a Hegelian and Marxian dialectical logic that stresses the analysis of antagonisms. Capurro's work is based on a thorough knowledge of, and engagement with, classical, modern and contemporary philosophy. Kant's philosophy has in this context been of particular relevance. Kant trusted that world peace could be achieved with the help of liberal democracy, world trade, and the political public (Capurro 2003, 78). Kant had the writing public in mind as the foundation for ethics and the Enlightenment. For Habermas, the communicating public is the foundation of ethics and politics.

Capurro stresses that the Internet, because of its own characteristics, cannot be a purely rational and enlightened space, but is one confronted by "semi-darkness" (Capurro 2003, 83). The questions about freedom of the press and freedom of speech would in the Internet age translate into the question about freedom of access (Capurro 2003, 79, 164, 198; 2006, 176). Capurro sees the United Nations as the best forum for discourses about Internet ethics (Capurro 2003, 80). He thereby argues for an institutional discursive form of Internet ethics. The moral values enshrined in the Universal Declaration of Human Rights are of central importance for information and Internet ethics (84, 199), specifically: human dignity, confidentiality, privacy, equality of opportunity, freedom of opinion and expression, participation in cultural life, and the protection of moral and material interests resulting from scientific, cultural, literary, and artistic production. Capurro stands with the foregrounding of human rights in Internet ethics in a Kantian tradition. This is expressed in his demand for a human right to freedom of communication on the Internet (137). One certainly must see how such freedoms remain in asymmetric societies class-structured. Economic and political power limits freedom so that universal ethical and legal claims are practically undermined and remain unrealised.

Capurro (1981) first used the term information ethics in 1981 and also grounded it in his habilitation thesis *Hermeneutik der Fachinformation* in 1986. This was ten years before Luciano Floridi, who has also used the term information ethics (Floridi 2013), published his first book, a book whose focus was not on ethics, but rather on epistemology. Similar to the tension between Manuel Castells and Jan van Dijk, the latter who invented the (nonsensical) term the network society, there remains a tension between Capurro and Floridi concerning the grounding of information ethics. Floridi (2013, 23) says that it "seems that information ethics began to merge with computer ethics only in the nineties". Capurro's (1986) treatment of information ethics in his habilitation definitely merges aspects of information and computer ethics earlier on. Floridi does not seem terribly willing to engage with approaches alternative to his own definitions for the field in any significant detail.[1] At the same time, one must say that Capurro's habilitation is also not generally accessible because it was only published in German, which limits international academic discourse. Floridi (2013, 19) finds it "unfortunate" that there are different versions of computer, information, and Internet ethics and says that his approach is "a unified approach". Floridi's unifying approach is not universalist enough because it requires a quite particularistic approach that is implicitly grounded in actor-network theory and post-humanist philosophy. It is, therefore, quite likely to attract criticism from other philosophers such as Capurro, who had already used the term information ethics before Floridi started doing so.

Floridi (2010) argues that information and communication technologies (ICTs) have brought about a revolution that resulted in an "informational turn" (11) that has been so profound that it has re-ontologised the world. The result would have been the emergence of a digitised infosphere, in which IT entities blur all boundaries and digitise all existence so that "*connected informational organisms (inforgs)*" come into existence (12). A new form of ethical constructionism would be needed that fights a "struggle against entropy" (17) and negotiates "a fruitful, symbiotic relationship between technology and nature" (18). Inforgs are, for Floridi, not just human. Therefore information ethics is for him

> "an environmental approach, one which does not privilege the natural or untouched, but treats as authentic and genuine all forms of existence and behaviour, even those based on artificial, synthetic, hybrid, and engineered artefacts. The task is to formulate an ethical framework that can treat the infosphere as a new environment worth the moral attention and care of the human inforgs".
>
> (Floridi 2013, 18)

Humans would be confronted with information resources that they use for creating information products that are immersed into and affect an information environment as a target (20). Information ethics therefore would have to reflect on moral issues concerning information resources, products, and targets. Floridi adds that his initial model (Floridi 2013, 20–21) is too limited at a micro-level and needs to be complemented by macroethics (25–28).

Information Ethics: Capurro and Floridi

Floridi's information ethics is non-, post-, and trans-humanist; it wants to be an ethics that considers all beings as actors in an informational environment:

"From an IE perspective, the ethical discourse now comes to concern information as such; that is, not just all persons, their cultivation, well-being, and social interactions, and not just animals, plants, and their proper natural life either, but also anything that may or will exist, like future generations; and anything that was, but is no more, like our ancestors. Unlike other non-standard ethics, IE is more impartial and universal – or one may say less ethically biased – because it brings to ultimate completion the process of enlarging the concept of what may count as a centre of moral claims, which now includes every instance of information, no matter whether physically implemented or not".

(Floridi 2013, 65)

Floridi's approach is pan-informational: he sees information everywhere, as a substance of the world. This becomes evident when he characterises the infosphere as "[m]aximally [...] a concept that, given an informational ontology, can also be used as synonymous with reality, or Being" (Floridi 2013, 6) or as "informational metaphysics" (307). Entropy is a crucial concept in Floridi's information ethics. Given that this concept tends to be used in thermodynamics as a measure of disorder and chaos and in Shannon's mathematical theory of communication as a measure of the uncertainty of information, Floridi (2013, 65) admits that the use of this notion in ethics can easily be misleading. He defines metaphysical entropy as "Non-Being", "absence or negation of any information" (65), "the decrease or decay of information leading to the absence of form, pattern, differentiation, or content in the infosphere" (67). Floridi formulates four information-ethical principles that apply to all actants and the totality of the infosphere:

- "entropy ought not to be caused in the infosphere (null law)",
- "entropy ought to be prevented in the infosphere",
- "entropy ought to be removed from the infosphere".
- "the flourishing of informational entities as well as the whole infosphere ought to be promoted by preserving, cultivating, and enriching their well-being" (Floridi 2013, 71).

Capurro argues that given the existing information overload, ever more information is not necessarily desirable because humans cannot handle it and it fragments their communication. Floridi's information-ethical entropy-reduction and -destruction programme would therefore be mistaken. "But do we not have enough information in the information society? It seems that this imperative would make the situation even worse than it is!" (Capurro 2003, 167). Capurro adds that Floridi's norms contradict "deleting viruses, SPAM and all kind of 'non useful' information" (Capurro 2008, 170). Floridi's information ethics is also problematic from a political perspective: assume we live in Nazi Germany in the years 1933–1945, a society dominated by anti-Semitic, racist, fascist, and imperialist ideology. This ideology has not ceased to exist after 1945. The principle of reducing metaphysical entropy implies that the presence of any ideology is good and that the more of it that is spread, the better. The real ethical imperative can however only be that Nazi ideology should be destroyed, i.e. informational entropy be increased because it is the worst imaginable system of domination and exploitation. Floridi understands the infosphere and information ethics as expansive so that all entities are subject to moral judgments. In these terms, one could define the Nazi regime as entropic because it sets out to annihilate Jews and political opponents – physically and thereby also their ideas. But what is the right answer to the Nazis? The only morally justified answer can be Adorno's "new categorical imperative" that humans "arrange their thoughts and actions so that Auschwitz will not repeat itself, so that nothing similar will happen" (Adorno 1973, 365).

In the situation of being inside Nazi Germany this then actually means that the ethical imperative must be to decrease homogeneity by increasing political entropy, i.e. by conducting anti-fascist attacks that aim to kill Hitler and other Nazi leaders and taking measures that aim to annihilate Nazi ideology. Destroying Nazism with violent and political means increases political entropy in order to enable a society that is not based on a project of extermination. Anti-fascist resistance is therefore in Floridi's terms the increase of political entropy. It aims at a society that does not systematically reduce entropy. Floridi's ethics cannot grasp these complexities. His ontology fails to provide an ethical imperative that can challenge Nazism and fascism. If any information is good, then also the ideologies of Nazism and fascism are good. Floridi argues that

> "because we have no reason against the intrinsic value of Being in all its manifestations, we should expand an environmental approach to all entities, including non-sentient beings. The injunction is to treat something as intrinsically valuable and hence worthy of moral respect by default, until 'proven guilty'".
>
> (Floridi 2013, 318)

The assumption that humans are or can be on one ontological level with non-human entities was proven wrong by Auschwitz. A biologistic and anti-Semitic ideology that describes groups of people as subhumans and parasites enabled Auschwitz. There are substantive historical reasons why we should refuse philosophies such as post-humanism, actor network theory, and Floridi's philosophy that argue that humans and non-humans are ontological equivalents.

For Floridi, companies, machines, or parties (Floridi 2013, 159) are also moral agents, which in his view is an assumption that holds the advantage of being "non-anthropocentric" (Floridi 2013, 58). Floridi positively acknowledges the non-anthropocentrism, or what some call "anti-speciesism" of deep ecology (133), and argues that his information ethics takes "this inclusive approach [...] further" (133). Floridi does not mention that critics of deep ecology have characterised versions of it as an eco-fascist movement (Bookchin 1987; Ditfurth 1996). Putting non-human beings onto the same moral level with humans, as both deep ecology and Floridi do, decentres human morality and affords an undifferentiated moral obligation to all living beings irrespective of origin. It is important to see how such approaches to decentring human morality are linked to strategies of exploitation in capitalism that reify human life: both treat human bodies and minds like things. Nazi ideology is an extreme form of reification. Strategies of exploitation in capitalism reify human beings: they treat their bodies and minds like things. The Nazis ideologically justified killing Jews by comparing them to parasites, which put humans on the same ethical level with animals. Anti-humanism is one of the first logical steps to fascism. Practical and ethical anti-fascism argues for the specificity and difference of the human being in relation to non-humans. This does not imply that humans should treat nature recklessly, but that the ethics of nature and the ethics of society have different qualities and principles.

Capurro (2008) argues, against Floridi's position, for a human-centred information ethics by stressing the difference between humans and things. Things-as-such would be morally worthless and humans "per se invaluable" (168). The value of things, such as their exchange value measured in money or their moral value associated with emotional attachment, arises out "of our relationship to others" (168). Only humans have the capacity to conduct economic evaluation (evaluating things) and moral evaluation ("evaluating ourselves") (169) and to relate both to each other. "As far as we know, we are the only living beings capable of mirroring the world as the common invaluable horizon that allows us to evaluate things" (169). Capurro (2008, 171) concludes his critique of Floridi by asking: "We have some 6 billion moral agents on earth. Why should we create millions (?) of artificial ones [to whom we assign 'moral responsibility']?". Capurro's human-centred ethics is not anthropocentric or individualistic, but social-relational. It asks us to "relativise our 'egocentric' ambitions" and poses the ethical question: "What is good for our bodily being-in-the world with others in particular?" (Capurro 2006, 182).

Floridi (2013, Chapter 14) conceives a business as an information process, in which the business provides, as an actor, goods or services to customers. He stresses that "profit is clearly not part of the essence of a business" (288) and that maximising profit is not a company's ethical imperative. Defining an economic organisation by orientation on exchange, profit, or money is indeed a crude form of fetishism that naturalises capital accumulation. A general definition of the economy is that it is a system, in which humans produce use-values that satisfy human needs. An economic organisation is an entity specialising in the production of specific use-values in order to satisfy human needs. Raymond Williams (1983, 79) points out in his *Keywords* that since the 15th century the English word "customer" has described "a buyer or purchaser". It is inevitably bound up with the modern forms of the market and capitalism. It is therefore inappropriate that Floridi uses the term customer when defining an economic organisation as "the provider of goods or services to customers" (Floridi 2013, 280). This formulation implies that markets, money, exchange value, and trade are inherent in all economies. The language often used in higher education systems that have been strongly commodified reveals the nature of this notion: students are often termed "customers" because they pay for (or rather go into debt, except if they have rich parents) access to education. The existence of online and offline gift economies, where people voluntarily give goods or services to others without the expectation of reciprocity or obtaining something in return, shows that trade is not an essential feature of the economy. A society of customers is a market and exchange society.

The three primary questions for information business ethics are for Floridi (2013, 284): "(1) What is provided? (2) How is it provided? (3) What impact does it have?". It is hard to see how the first two questions relate to ethics, whereas the third one can be related to ethics if one asks how the economy and economic organisations can have positive impacts that benefit all. The imperative for Floridi's information business ethics is fostering "human flourishing and avoiding wastefulness". He understands wastefulness as "*destruction, corruption, pollution*, and *depletion* of (parts of) reality" (290). Ecological problems are related to the mode of economic production, but are not the economy's only ethical dimension. It is difficult to frame exploitation – the main ethical social problem of all class societies – in terms of waste and entropy. It is worth highlighting that Floridi's analysis does not problematise exploitation. His information ethics does not give importance to the phenomena of class and exploitation and is, therefore, particularly unsuited for a critique of exploitation in the information age. For Marx (1867), exploitation means that one class whose labour produces use-values is deprived and excluded from them by another class that takes private ownership of these use-values, for the purposes of facilitating exchange and accumulation. The producing class is deprived of wealth and the owning class increases its wealth. Exploitation is a question of distributive justice and ownership justice, not one of waste, order, and disorder. The ethical social imperative for a critical theory of the economy and society is therefore that one needs to "overthrow all relations in which man is a debased, enslaved, forsaken, despicable being" (Marx 1844, 182).

Humans cannot exist without, and only exist in and through, social relations. Society is social-relational; it is based on human co-operation (Fuchs 2008). There can be no society without relations, communication, and co-operation. But a society without competition, war, markets, egoism, and exchange is perfectly possible (Fuchs 2008). Exploitation and domination limit our capacities to fully organise society by giving particularistic advantages to one group or individual over others. The ethical imperative is therefore to question and undo exploitation and domination and to create conditions that benefit all, i.e. a classless society without exploitation and domination.

Marxist political economy of information and communication is based on an inherently ethical imperative: it "goes beyond technical issues of efficiency to engage with basic moral questions of justice, equity and the public good" that concern information and communication (Murdock and Golding 2005, 61). The "moral dimension remains strong in Marxian political economy because it provides a powerful defence of democracy, equality, and the public sphere in the face of dominant private interests" (Mosco 2009, 34). Critical political economy of information and communication therefore analyses "the power relations, that mutually constitute the production, distribution, and consumption of resources, including communication resources" (Mosco 2009, 2).

4.2 Information Ethics in the Age of Digital Labour and Edward Snowden

I do not see myself as a representative of computer, information, digital media, or Internet ethics, but am rather interested in a critical theory and critique of the political economy of information, communication, technology, the media, and the Internet. Such an approach aims to theorise these phenomena's political economy and their power structures, to empirically analyse human realities within such structures, to conduct ideology critique of reifications of information, and to inform social struggles for alternatives. Ethics is one of the dimensions of this approach, but not its exclusive one. It also requires social theory and empirical social research. In this section, I want to discuss two information-ethical problems: digital labour and Internet surveillance.

4.2.1 Digital Labour

The production of information and information technology is embedded into an international division of information labour (Fuchs 2014, 2015). There are new technologies, but capitalism, imperialism, class, and exploitation continue to form the heart of society and international relations and shape the modes of information production, distribution, and consumption that have become so important in the 21st century. Critical scholars introduced the notion of the new international division of labour (NIDL) in the 1980s in order to stress that developing countries had become cheap sources of manufacturing labour and to highlight the rise of transnational corporations (TNCs) (Fröbel, Heinrichs and Kreye 1981). "Digital labour" is not a term that only describes the production of digital content. It is a category that rather encompasses the whole mode of digital production that contains a network of agricultural, industrial, and informational forms of work that enables the existence and usage of digital media. The international division of digital labour (IDDL) is the NIDL in the context of the production and productive use of digital media. The IDDL is a complex network that involves global interconnected processes of exploitation, such as the exploitation of Congolese slave-miners who extract minerals that are used as the physical foundation for ICT components that are manufactured by millions of highly exploited Fordist wage-workers in factories such as Foxconn, low-paid software engineers in India, highly paid and highly stressed software engineers at Google and other Western software and Internet corporations, or precarious freelancers in the world's global cities who are using digital technologies to create and disseminate culture, poisoned eWaste workers who disassemble ICTs and thereby come in touch with toxic materials, etc. (Fuchs 2014, 2015). Let us have a look at two forms of labour involved in the IDDL: mining of ICT-related minerals in the Congo and hardware assemblage in China.

Capitalism as the dominant mode of economic activity has not brought older modes of production to an end, but has rather subsumed them. Slavery and patriarchy continue to exist and to be modes of organisation for the super-exploitation of labour. In 2014, 35.8 million people lived in modern forms of slavery (Global Slavery Index 2014). Modern slavery includes slavery, debt bondage, forced marriage, sale and exploitation of children, forced labour, and human trafficking (ibid.). Slaves in the Democratic Republic of Congo mine a specific portion of the minerals (such as cobalt, coltan, and tin) needed for creating electronics and computing equipment (Fuchs 2014, Chapter 6). In 2014, the DRC was ranked 186 out of 187 countries in human development; 87.8% lived in extreme poverty on less than US$1.25 per day, and 38.8% of the population aged 15 or older was illiterate.[2] A combination of civil war and neo-imperialist exploitation of labour and the country's resources (that do not benefit local people, but primarily Western companies) has created the paradox – typical for capitalism – that one of world's richest countries in natural resources is socially the world's poorest country. In 2014, the political situation in the Democratic Republic of Congo (DRC) saw continued hostilities involving government forces, rebels, and fighters from Uganda and Rwanda. The country's inhabitants experienced war crimes, crimes against humanity, forced recruitment of children as soldiers, mass rapes, and the killing, mutilation, and torture of civilians.[3] According to estimations, more than 760,000 people in the DRC were slaves in 2014.[4] Following Nigeria, it is the country with the second-largest absolute number of slaves.

Apple was, according to the Forbes 2000 list of the largest transnational companies, the world's 15th largest company in 2014.[5] Its profits were US$37 billion in 2013 and 39.5 billion in 2014.[6] In 2014, iPhones accounted for 56% of Apple's net sales, iPads for 17%, Macs for 13%; iTunes, software, and services for 10%.[7] According to calculations published by Chan, Pun, and Selden (2013, 107), the Chinese labour involved in manufacturing an iPhone makes up only 1.8% of the iPhone's price, while Apple's profit margins are 58.5% and Apple's suppliers, such as the Taiwanese company Hon Hai Precision that is also known as Foxconn, account for 14.3% of revenues. Applying this information shows that the iPhone 6 Plus does not cost US$299 because of labour costs, but rather because Apple on average earns US$175 profits, Foxconn US$43 profits, and the workers assembling the phone in a Foxconn factory in total US$5. The high costs are a consequence of a high-profit rate and a high rate of exploitation that are achieved by organising digital labour within an international division. According to the CNN Global 500 2012 list,[8] Foxconn is the fifth largest corporate employer in the world. In 2011, Foxconn had enlarged its Chinese workforce to a million, a majority being young migrant workers coming from the countryside (SACOM 2011). Foxconn assembles the iPad, iMac, iPhone, the Amazon Kindle, and various consoles (by Sony, Nintendo, Microsoft). When 17 Foxconn workers attempted to commit suicide between January and August 2010 (most of them succeeded), the topic of bad working conditions in the Chinese ICT assemblage industry became widely known. This circumstance was followed up with a number of academic works that showed that workers' everyday reality at Foxconn includes low wages, working long hours, frequent work shift changes, regular working time of over ten hours per day, a lack of breaks, monotonous work, physical harm caused by chemicals such as benzene or solder paste, lack of protective gear and equipment, forced use of students from vocational schools as interns (in agreement with the school boards) that conduct regular assembly work that does not help their studies, prison-like accommodations with 6–22 workers per room, yellow unions that are managed by company officials and whom the workers do not trust, harsh management methods, a lack of breaks, prohibitions that workers move, talk or stretch their bodies, the requirements that workers stand during production, punishments, beatings and harassments by security guards and disgusting food (Chan 2013; Chan, Pun and Selden 2013; Fuchs 2014, Chapter 7; Pun and Chan 2012; Qiu 2012; Sandoval 2013).

Apple claims in its *Supplier Responsibility 2014 Progress Report* that it drove its "suppliers to achieve an average of 95 percent compliance with our maximum 60-hour work week".[9] The International Labour Organization's Convention C030 – Hours of Work recommends that the working week should not last longer than 48 hours. That Apple defines itself a standard of 60 hours for labour in its supply chain and prides itself for this fact shows that imperialism's international division of labour is not just exploitative, but also racist in character: Apple assumes that for people in China, 60 hours is an appropriate standard of working time. Apple's argument is based on the Western assumption that Asians have a strong work ethic and are therefore suited to work long hours for comparatively low wages. It undermines the universal assumptions enshrined in the ILO Convention that there is a maximum of hours that human labour should not exceed because otherwise life is reduced to labour time.

Apple says that for its 2014 report it audited the working conditions of more than 1 million workers. It is however a fact that these audits are not conducted independently and that the results are also not reported independently. Apple doesn't rely on independent corporate watchdog organisations such as Students & Scholars against Corporate Misbehaviour (SACOM), but rather conducts studies that one can only consider to be biased. Workers who are studied by their own employers will certainly not report what they think is wrong because they are afraid to lose their job. Apple's report is written in a style and language that conveys the impression that suppliers and local agencies that behave immorally are the problem: "Our suppliers are required to uphold the rigorous standards of Apple's Supplier Code of Conduct, and every year we raise the bar on what we expect. [...] We audit all final assembly suppliers every year". That such behaviour is however driven by TNCs' demand to produce cheaply and quickly is never mentioned. Apple uses the ideological strategy that it emphasises positive things about itself and negative things about suppliers in order to distort attention from its own responsibility for the exploitation of Chinese workers. In 2014, SACOM published a new report on working conditions at Apple's supplier Pegatron in Jiangsu,[10] where tens of millions of the iPhone 6 have been manufactured. Undercover scholars conducted the research.

Information Ethics in the Age of Digital Labour and Edward Snowden

"Workers told SACOM researchers that they sometimes have to work very long hours till early morning, often 12 to 15 hours a day, and sometimes even up to 17 to 18 hours a day. In other words, the total amount of overtime hours can be up to 170 to 200 hours a month, which, in turn, means that workers have to work more than 360 hours a month".

(SACOM 2014, 2)

Further issues at Pegatron included an unsafe and unhealthy working environment, illegal charges for health checks, insufficient health information, precarious dispatch labour, exclusion from social insurance, difficulties to resign from the job, scolding, fines, repressive management, and lack of trade unions. The report concludes:

"Pegatron and its buyer Apple have continuously engaged in poor labour practices and abuses of workers' rights. Even though the Apple Inc. has established its code of conducts since 2005, the working conditions in Apple's supply chain are still far from satisfactory. This report, along with the earlier investigative reports released by SACOM throughout the years, have continuously demonstrated that Apple and its suppliers in the Chinese mainland have never treated their workers with dignity".

(SACOM 2014, 21)

A 2014 BBC undercover report unveiled that workers assembling iPhones 6 in Pegatron factories are so overworked that they fall asleep during work and in their breaks.[11]

An ideology is a claim that does not correspond to and tries to distort the representation of reality. SACOM's studies show that reality in the factories of Apple's Chinese suppliers is different than reported in the company's own reports. Apple tries to distort presentations of labour in its supply chain by ideology in order to forestall critique of capitalism. Why is the exploitation of digital labour, for which the Congo and the Foxconn cases are good examples, ethically problematic? Capitalistically produced digital media are not accessible for all people in the world and not to the same extent and with the same benefits. The benefits of the one, especially digital media companies that derive large monetary benefits from selling hardware, software, content, access, audiences, and users, stem from the misery of the labour of others. There is not just a power asymmetry immanent in the IDDL, but a fundamental injustice that creates conditions that deprive digital workers of their humanity, make them work under conditions not adequate for any human being, and result in distributive injustice so that the benefits from digital media are asymmetrically distributed so that the class of digital capitalists enriches itself by depriving others. Let us go back to two fundamental questions that Capurro's information ethics ask: "How can we ensure that the benefits of information technology are not only distributed equitably, but that they can also be used by the people to shape their own lives?" (Capurro 2003, 41) and "What is good for our bodily being-in-the world with others in particular?" (Capurro 2006, 182). The problem of the capitalist mode of organising digital media, i.e. the IDDL, is that it creates distributive injustice. It only enables some people to use these media to shape their own lives. It results in conditions of slavery and exploitation, in which humans cannot determine their own lives and cannot own the products their life activities create. It constitutes a being-in-the-world with others, where one class appropriates the labour and products of digital workers. It thereby creates inverse interdependent welfare (Wright 1997) for itself coupled with the deprivation of opportunities for others and their exclusion from this appropriated welfare.

4.2.2 Internet Surveillance in the Age of Edward Snowden

In June 2013, Edward Snowden revealed with the help of the *Guardian* the existence of large-scale Internet and communications surveillance systems such as Prism, XKeyscore, and Tempora. According to the documents he leaked, the National Security Agency (NSA), through the Prism programme, obtained direct access to user data from seven online/ICT companies: AOL, Apple, Facebook, Google, Microsoft, Paltalk, Skype, and Yahoo![12] The Powerpoint slides that Edward Snowden leaked refer to data collection "directly from the servers of these U.S. Service Providers"[13] (ibid.). Snowden also revealed the existence of a surveillance system called XKeyScore that the NSA can use for reading e-mails, tracking web browsing and users' browsing histories, monitoring social media activity, online searches, online chat, phone calls, and online contact networks, and follow the screens of individual computers. According to the leaked documents, XKeyScore can search both meta-data and content data.[14]

The documents that Snowden leaked also showed that the Government Communications Headquarter (GCHQ), a British intelligence agency, monitored and collected communication phone and Internet data from fibre optic cables and shared such data with the NSA.[15] According to the leak, the GCHQ, for example, stores phone calls, e-mails, Facebook postings, and the history of users' website access for up to 30 days and analyses these data.[16] Further documents indicated that in co-ordination with the GCHQ, intelligence services in Germany (Bundesnachrichtendienst BND), France (*Direction Générale de la Sécurité Extérieure* DGSE), Spain (Centro Nacional de Inteligencia, CNI), and Sweden (Försvarets radioanstalt FRA) developed similar capacities.[17]

Edward Snowden's revelations about the existence of surveillance systems such as Prism, XKeyScore, and Tempora have shed new light on the extension and intensity of state institutions' Internet and social media surveillance. The concept of the military-industrial complex stresses the existence of collaborations between private corporations and the state's institutions of internal and external defence in the security realm. C. Wright Mills argued in 1956 that there is a power elite that connects economic, political, and military power:

> "There is no longer, on the one hand, an economy, and, on the other hand, a political order containing a military establishment unimportant to politics and to money-making. There is a political economy linked, in a thousand ways, with military institutions and decisions. [...] there is an ever-increasing interlocking of economic, military, and political structures".
>
> (Mills 1956, 7–8)

Edward Snowden has confirmed that the military-industrial complex contains a surveillance-industrial complex (Hayes 2012), into which social media are entangled: Facebook and Google each have more than 1 billion users and have likely amassed the largest collection of personal data in the world. They and other private social media companies are first and foremost advertising companies that appropriate and commodify data on users' interests, communications, locations, online behaviour, and social networks. They make a profit out of data that users' online activities generate. They continuously monitor usage behaviour for this economic purpose. Since 9/11 there has been a massive intensification and extension of surveillance that is based on the naïve technological-deterministic surveillance ideology that monitoring technologies, big data analysis, and predictive algorithms can prevent terrorism. The reality of the murder of a soldier that took place in the South-East London district of Woolwich in May 2013 and the Charlie Hebdo attacks in Paris in January 2015 shows that terrorists can use low-tech tools such as machetes and conventional guns for targeted killings. High-tech surveillance will never be able to stop terrorism because most terrorists are smart enough not to announce their intentions on the Internet. It is precisely this surveillance ideology that has created intelligence agencies' interest in the big data held by social media corporations. Evidence has shown that social media surveillance not just targets terrorists, but has also been directed at protestors and civil society activists.[18] State institutions and private corporations have long collaborated in intelligence, but the access to social media has taken the surveillance-industrial complex to a new dimension: it is now possible to obtain detailed access to a multitude of citizens' activities in converging social roles conducted in converging social spaces.

The profits made by social media corporations are not the only economic dimension of the contemporary surveillance-industrial complex: the NSA has subcontracted and outsourced surveillance tasks to approximately 2000 private security companies[19] that make profits by spying on citizens. Booz Allen Hamilton, the private security company that Edward Snowden worked for, is just one of these firms that follow the strategy of accumulation-by-surveillance. According to financial data,[20] it had 24,500 employees in 2012 and its profits increased from US$25 million in 2010 to 84 million in 2011, 239 million in 2012, 219 million in 2013, and 232 million in 2014. Surveillance is big business, both for online companies and those conducting the online spying for intelligence agencies.

Users create data on the Internet that is either private, semi-public, or public. In the social media surveillance-industrial complex, companies commodify and privatise user data as private property, and secret services such as the NSA driven by a techno-determinist ideology obtain access to the same data for trying to catch terrorists that may never use these technologies for planning attacks. For organising surveillance, the state makes use of private security companies that derive profits from organising the monitoring process.

User data is in the surveillance-industrial complex first externalised and made public or semi-public on the Internet in order to enable users' communication processes, then privatised as private property by Internet platforms in order to accumulate capital, and finally particularised by secret services that bring massive amounts of data under their control, data that is made accessible and analysed worldwide with the help of profit-making security companies. Why is the surveillance-industrial complex problematic from an ethical point of view? Let us again have a look at the foundational questions of Capurro's information ethics. "How can we ensure that the benefits of information technology are not only distributed equitably, but that they can also be used by the people to shape their own lives?" (Capurro 2003, 41). "What is good for our bodily being-in-the world with others in particular?" (Capurro 2006, 182). The surveillance-industrial complex contains fundamental power asymmetries: the involved nation states argue that they have to monitor the communication of all citizens worldwide beyond nation states, but at the same time they want to hinder citizens monitoring state power, as the repression against WikiLeaks, Chelsea Manning, and Edward Snowden shows.

Information Ethics in the Age of Digital Labour and Edward Snowden

The surveillance-industrial complex is also asymmetrical in terms of knowledge because it wants to deceive the world by not making transparent the existence of global surveillance systems. This is a strategy frequently found in surveillance that is hard to criticise because it operates invisibly and covertly. Unknown powers can hardly be questioned. The world should be grateful to Edward Snowden and award him and Julian Assange the Nobel Peace Prize for having made many unknowns known so that global society has become better enabled to criticise the existing power elites that operated behind their backs. Surveillance, as the collection of information about people in order to enforce power structures within a society, is not automatically a bad thing. If a government or civic watchdog for example monitors financial flows and corporate power in order to uncover and overcome corporate crime and corporate tax evasion in order to increase the public tax revenues, then the use of the surveillance power strengthens the public good. At a macro-level, this form of surveillance therefore benefits society at large. Within a society that is based on asymmetric power structures, not all forms of surveillance are morally problematic. The surveillance-industrial complex that Snowden exposed is morally problematic because it is based on the economic exploitation of digital labour, the deception of the public, a power asymmetry that tries to repressively block watchdogs' monitoring of state and corporate power, and surveillance ideologies that create the false impression that more surveillance results in more security and solves political and social problems.

The information technologies of the surveillance-industrial-complex disempower citizens who cannot shape their own conditions of information and it creates multiple power asymmetries that question the freedoms of information, thought, opinion, and communication that liberal societies claim as their fundamental moral values. The surveillance-industrial complex shows that a negative dialectic of the Enlightenment is at play in contemporary society: it constantly undermines the very liberal values of the Enlightenment, such as the freedoms of thought, speech, press, and assembly as well as the security of the person and of their personal property. Prism shows how in supposedly liberal democracies dangerous forms of political-economic power negate Enlightenment values (Fuchs 2015, Chapter 8).

Surveillance ideologies – such as "if you got nothing to hide, then you got nothing to fear", "for security we need to compromise some privacy", "surveillance will stop crime and terrorism" – are mistaken for many reasons:

- Terrorists are not so silly as to communicate online what they are doing or intend to do.
- There is no technological fix to political and socio-economic problems.

- Law and order politics fosters fascist potentials in society.
- Categorical suspicion turns the presumption of innocence ("innocent until proven guilty") into a presumption of guilt ("terrorist until proven innocent").
- People who join fundamentalist groups often experience precarity, unemployment, lack of good educational opportunities, and racism. Welfare state politics, not politics of control, are the best means for countering fundamentalism.

Times of crisis are times of ideological scapegoating in order to distract attention from the causes of social problems. In 2008, a major crisis of capitalism started. It also translated into a crisis of many states and societies. The emergence of heavy ideological scapegoating is therefore no surprise. Contemporary scapegoats in the UK context include Romanian and Bulgarian workers, the European Union, benefits recipients, the unemployed, the poor, black youth, international students, immigrants, Muslims, Jews, South Europeans, etc. Ideology deflects attention from social problems, inequality, precarious labour, and unemployment. It deflects attention from the problems of capitalism.

Moral panics that call for more surveillance and scapegoat certain groups can amplify and result in more terrorism and crime: if groups or individuals feel unfairly discriminated (e.g. by racism, classism, sexism, scapegoating, etc.), they may react to this circumstance with an intensification of hatred against those whom they perceive hate and discriminate against them. If certain groups or individuals are labelled as terrorists or criminals or denied certain possibilities (such as entering a certain country, area, or building), there is the risk that an intensification or creation of hate can set in, which can result in the creation or intensification of the very phenomenon (crime, terror, etc.) that the algorithm, surveillance technology, ideology, law and order policy, etc. wanted to prevent in the first instance. The European protests and rejections of austerity, neoliberalism, and capitalism are, in my view, the only reasonable voices in the crisis discourse. Slavoj Žižek (2015) pinpoints this circumstance by saying that a "renewed Left" is "the only way to defeat fundamentalism, to sweep the ground under its feet". Syriza's electoral victory in Greece is an important beacon of hope for the Left in Europe, a hope for a world beyond ideology, right-wing populism, and neoliberalism.

4.3 Conclusion

A critical theory and critical political economy of information, communication, technology, the media, and the Internet need to be a theoretical, empirical, ethical, and political inquiry into the information society's power structures It must also uncover, question, and help to overcome the inequality, power asymmetries, exploitation, ideologies, and forms of domination that emerge in the context of information and information technologies. The question therefore arises of how information ethics should best be conceived. I have analysed in this contribution the relationship between two versions of information ethics, the ones formulated by Rafael Capurro and Luciano Floridi. Floridi's approach is highly problematic because it decentres the human and thereby risks relativising the very foundations of ethics. He does not engage with the critiques of deep ecology, post-humanism, and actor-network theory that face the same problems as his version of information ethics. Floridi (2013, 308) argues about a specific claim that once was made against him is:

> "I still recall one conference in the nineties when a famous computer ethicist compared me to a sort of Nazi, who wished to reduce humans to numbers, pointing out that the Nazis used to tattoo six-digit identity tags onto the left arms of the prisoners in their Lager. This is rhetorical nonsense".

Although this is certainly an overdrawn claim, Floridi simply dismisses it and does not ask himself if there may be certain problematic assumptions at the heart of his philosophy that make some people feel politically uncomfortable and make them think that it trivialises the horrors of Nazism.

Floridi overlooks that biologism, as an ideology that equates humans and non-humans by arguing that certain humans are like parasites or other biological organisms, or by projecting biological mechanisms into society, is one of the important logical foundations of Nazism. It makes it logically possible to treat humans like things and to ideologically argue that they do not deserve to exist. Floridi certainly can reject this line of argument because he argues that all existence is informational and is valuable and should not be destroyed. This however also implies that not just computer viruses, but also SARS-CoV-2, the human immunodeficiency virus (HIV), and other viral illnesses that can threaten human lives should be preserved, which means the death of humans. Such assumptions in some versions of deep ecology threaten human lives and have resulted in a form of eco-fascism. Floridi does not engage with such approaches and their problems. The point is that an ontological equalisation of humans and non-humans has historically been the foundation of repression and that ontological equalisations as right-wing ideology continues to exist for example in the ideology of some animal rights activists and the deep ecology movement. The social-ecological philosopher Murray Bookchin (1987) warned in this context

> "Deep ecology contains no history of the emergence of society out of nature [...] 'Biocentric democracy,' I assume, should call for nothing less than a hands-off

policy on the AIDS virus and perhaps equally lethal pathogens that appear in the human species. [...] Deep ecology, with its Malthusian thrust, its various centricities, its mystifying Eco-la-la, and its disorienting eclecticism degrades this enterprise into a crude biologism that deflects us from the social problems that underpin the ecological ones and the project of social reconstruction that alone can spare the biosphere from virtual destruction".

Floridi's information ethics faces the danger of reproducing some of the problems of deep ecology.

Rafael Capurro has, in contrast to Floridi, grounded a form of information ethics that foregrounds human social relations as constitutive for the ethical understanding of information technologies and society. One can well disagree with Capurro on how to assess Heidegger, Kant, Vattimo, Marx, Hegel, etc., but in terms of the bottom line it is clear that his ethics cares about the deconstruction of asymmetric power structures and ideologies, which is a good foundation for constructive agreement and disagreement with political economy approaches. Floridi's pan-informational ethics foregrounds the reduction of entropy and the centrality of human and non-human actors that are conceived as having in common the simple quality that they are merely informational. It also stresses the struggle against all beings' entropy. It is hard to find this approach fruitful if one wants to develop a critical theory and critical political economy of information, the information society, and information technology.

Information ethics is an important philosophical undertaking that we require for a better understanding of the 21st century. We require, however, not just a general understanding, but specifically a critical understanding of the information society. Rafael Capurro's works are an important and indispensable contribution towards this philosophical task.

Notes

1 There is for example only one brief clause mentioning Capurro in Floridi's (2013, 308) book *The Ethics of Information*, whereas Capurro (2006, 2008) has published two major articles dedicated entirely to the discussion of Floridi's work.
2 Data source: Human Development Indicators, http://hdr.undp.org/ (accessed on 26 October 2014).
3 Source: Human Rights Watch 2014 Report: Democratic Republic of Congo: http://www.hrw.org/world-report/2014/country-chapters/democratic-republic-congo (accessed on 22 December 2014).

4 Ibid.

5 http://www.forbes.com/global2000/list (accessed on 22 December 2014).

6 Apple SEC filings, form 10-K, 2014.

7 Ibid.

8 http://money.cnn.com/magazines/fortune/global500/2012/full_list/ (accessed on 29 October 2013).

9 https://www.apple.com/supplier-responsibility/pdf/Apple_SR_2014_Progress_Report.pdf (accessed on 22 December 2014).

10 See also the 2013 investigation by China Labor Watch: http://www.chinalaborwatch.org/report/68. A comparable case is the iPhone 6 assemblage at Jabi in Wuxi: http://www.chinalaborwatch.org/report/103

11 Apple 'failing to protect Chinese factory workers'. http://www.bbc.com/news/business-30532463 *BBC Online*, December 18, 2014.

12 NSA Prism program taps in to user data of Apple, Google and others. *The Guardian Online*. June 7, 2013. http://www.theguardian.com/world/2013/jun/06/us-tech-giants-nsa-data

13 Ibid.

14 XKeyscore: NSA tool collects "nearly everything a user does on the internet'. *The Guardian Online*. July 31, 2013. http://www.theguardian.com/world/2013/jul/31/nsa-top-secret-program-online-data

15 GCHQ taps fibre-optic cables for secret access to world's communications. *The Guardian Online*. June 21, 2013. http://www.theguardian.com/uk/2013/jun/21/gchq-cables-secret-world-communications-nsa?guni=Article:in%20body%20link

16 Ibid.

17 GCHQ and European spy agencies worked together on mass surveillance. The Guardian Online. November 1, 2013. http://www.theguardian.com/uk-news/2013/nov/01/gchq-europe-spy-agencies-mass-surveillance-snowden

18 Spying on Occupy activists. *The Progressive Online*. June 2013. http://progressive.org/spying-on-ccupy-activists

19 A hidden world, growing beyond control. *Washington Post Online*. http://projects.washingtonpost.com/top-secret-america/articles/a-hidden-world-growing-beyond-control/

20 SEC Filings, http://investors.boozallen.com/sec.cfm).

References

Adorno, Theodor. W. 1973. *Negative Dialectics*. London: Routledge.

Bookchin, Murray. 1987. Social Ecology versus Deep Ecology: A Challenge for the Ecology Movement. *Green Perspectives*, 4–5, 1–23. Online version: http://dwardmac.pitzer.edu/Anarchist_Archives/bookchin/socecovdeepeco.html

Capurro, Rafael. 2008. On Floridi's Metaphysical Foundation of Information Ecology. *Ethics and Information Technology* 10 (2–3): 167–173.

Capurro, Rafael. 2006. Towards an Ontological Foundation of Information Ethics. *Ethics and Information Technology* 8 (4): 175–186.

Capurro, Rafael. 2003. *Ethik im Netz*. Stuttgart: Franz Steiner Verlag.

Capurro, Rafael. 1986. *Hermeneutik der Fachinformation*. Freiburg: Alber.

Capurro, Rafael. 1981. Zur Frage der Ethik in Fachinformation und –kommunikation. *Nachrichten für Dokumentation* 32 (1): 9–12.

Chan, Jenny. 2013. A Suicide Survivor: The Life of a Chinese Worker. *New Technology, Work and Employment* 28 (2): 84–99.

Chan, Jenny, Ngai Pun, and Mark Selden. 2013. The Politics of Global Production: Apple, Foxconn and China's New Working Class. *New Technology, Work and Employment* 28 (2): 100–115.

Ditfurth, Jutta. 1996. *Entspannt in die Barbarei. Esoterik, (Öko-)Faschismus und Biozentrismus.* Hamburg: Konkret.

Floridi, Luciano. 2013. *The Ethics of Information*. Oxford: Oxford University Press.

Floridi, Luciano. 2010. Information ethics. In *The Cambridge Handbook of Information and Computer Ethics*, ed. Luciano Floridi, 77–97. Cambridge: Cambridge University Press.

Fröbel, Folker, Jürgen Heinrichs, and Otto Kreye. 1981. *The New International Division of Labour.* Cambridge: Cambridge University Press.

Fuchs, Christian. 2015. *Culture and Economy in the Age of Social Media*. New York: Routledge.

Fuchs, Christian. 2014. *Digital Labour and Karl Marx*. New York: Routledge.

Fuchs, Christian. 2008. *Internet and Society. Social Theory in the Information Age.* New York: Routledge.

Global Slavery Index. 2014. *The Global Slavery Index 2014*. Dalkeith (Western Australia): Walk Free Foundation.

Hayes, Ben. 2012. The Surveillance-Industrial Complex. In *Routledge Handbook of Surveillance Studies*, ed. Kirstie Ball, Kevin D. Haggerty and Dadvid Lyon, 167–175. Abingdon: Routledge.

Marx, Karl. 1867. *Capital. Volume 1*. London: Penguin.

Marx, Karl. 1844. Introduction to the Critique of Hegel's Philosophy of Law. In *Marx & Engels Collected Works (MECW) Volume 3*, 175–187. New York: International Publishers.

Mills, C. Wright. 1956. *The Power Elite*. Oxford: Oxford University Press.

Mosco, Vincent. 2009. *The Political Economy of Communication*. London: Sage.

Murdock, Graham and Peter Golding. 2005. Culture, Communications and Political Economy. In *Mass Media and Society*, ed. James Curran and Michael Gurevitch, 60–83. London: Hodder Arnold.

Pun, Ngai and Jenny Chan. 2012. Global Capital, the State, and Chinese Workers: The Foxconn Experience. *Modern China* 38 (4): 383–410.

References

Qiu, Jack L. 2012. Network Labor: Beyond the Shadow of Foxconn. In *Studying Mobile Media: Cultural Technologies, Mobile Communication, and the iPhone*, ed. Larissa Hjorth, Jean Burgess and Ingrid Richardson, 173–189. New York: Routledge.

Sandoval, Marisol. 2013. Foxconned Labour as the Dark Side of the Information Age: Working Conditions at Apple's Contract Manufacturers in China. *tripleC: Communication, Capitalism & Critique* 11 (2): 318–347.

Students & Scholars Against Corporate Misbehaviour (SACOM). 2014. *The Lives of iSlaves: Report on Working Conditions at Apple's Supplier Pegatron*. http://sacom.hk/wp-content/uploads/2018/10/2014-The-Lives-of-Apple%E2%80%99s-iSlave-in-its-Pegatron-Sweatshops-in-China.pdf

Students & Scholars Against Corporate Misbehaviour (SACOM). 2011. *iSlave Behind the iPhone. Foxconn Workers in Central China*. https://www.somo.nl/wp-content/uploads/2011/10/iSlave-behind-the-iPhone-Foxconn-Workers-in-Central-China.pdf

Williams, Raymond. 1983. *Keywords. A Vocabulary of Culture and Society*. London: Fontana. Revised edition.

Wright, Erik Olin. 1997. *Class Counts: Comparative Studies in Class Analysis*. Cambridge: Cambridge University Press.

Žižek, Slavoj. 2015. Slavoj Žižek on the Charlie Hebdo Massacre: Are the Worst Really Full of Passionate Intensity? *New Statesman Online*, 10 January 2015: http://www.newstatesman.com/world-affairs/2015/01/slavoj-i-ek-charlie-hebdo-massacre-are-worst-really-full-passionate-intensity

Chapter Five

"Dear Mr. Neo-Nazi, can you please give me your informed consent so that I can quote your fascist tweet?": Questions of Social Media Research Ethics in Online Ideology Critique

---

5.1 Introduction

Consider the following tweets posted on 9 November 2016, one day after Donald Trump won the US presidential election:

> "President **Trump** wants to know if you have any last words Mr Soros?" #RevengeWillBeSweet #WhiteGenocide #RapeJIhad #RWDS #Trump #Trump16 [+ image of a Nazi shooting a Jewish person]"

> "#Trump 卐 The end of #WhiteGenocide in America. #Nazi #SiegHeil"

> "We won! This is a BIG win for the white race as a whole. And we won't stop. We will take back what is ours! #MAGA #WhitePride #14words"

> "Anti-Whites are shitting themselves right now. They do not like whites taking back their country!! #WhitePride #Trump2016"

> "Gonna go kill some **niggers,** mexicans, and muslims tommorow **trump** will just pardon me lol cant wait wooo #MAGA"

The examples indicate the prevalence of fascist, racist, and nationalist ideology in public discussions of Trump's victory. Given that the world economic crisis of 2008 has turned into a political crisis that has brought about the intensification of nationalism, xenophobia, racism, and fascism, it is an important task for critical research to study how and why these phenomena exist. Social media is a kind of mirror of what is happening

DOI: 10.4324/9781003279488-7

in society. Studying social media content is therefore a good way of studying society. But whenever we conduct social research, ethical issues regarding anonymity, informed consent, and privacy may arise.

Research ethics is a key aspect of social science. Not only is there a general etiquette of publishing, but also ethical questions that arise in the collection of data. The emergence of what some call "social media" and "big data" has complicated research ethics. In this contribution, I reflect on research ethics in respect to the study of online ideologies, especially in the context of "negative" social movements and forms of online expression that are fascist, racist, nationalist, anti-socialist, and anti-Semitic in character.

Doing online research complicates research ethics. So when for example conducting a Critical Discourse Analysis (CDA) of white supremacist content, the question arises whether you have to obtain informed consent for including and analysing a fascist tweet. Writing an e-mail asking, "Dear Mr. Neo-Nazi, can you please give me your informed consent so that I can quote your fascist tweet?", may not just result in rejection, it could also draw the attention of fascists towards you as a critical researcher and put you in danger.

This chapter deals with the question of how to deal with research ethics in qualitative online research. First, the chapter discusses the limits of established research ethics guidelines (Section 5.2). Second, it outlines the foundations of critical-realist Internet research ethics (Section 5.3). Third, it provides some examples of how to use such a framework (Section 5.4). Finally, some conclusions are drawn (Section 5.5).

5.2 Established Research Ethics Guidelines

An obvious approach of how to deal with questions of research ethics in qualitative online research is to look at established research ethics guidelines provided by academic associations.

The Association of Internet Researchers' ethical recommendations (2012) contain a list of questions that one can ask when conducting online research and points out ethical problems that may arise:

> "People may operate in public spaces but maintain strong perceptions or expectations of privacy. Or, they may acknowledge that the substance of their communication is public, but that the specific context in which it appears implies

restrictions on how that information is – or ought to be – used by other parties. Data aggregators or search tools make information accessible to a wider public than what might have been originally intended. [...] Social, academic, or regulatory delineations of public and private as a clearly recognizable binary no longer holds in everyday practice. [...] Yet there is considerable evidence that even 'anonymised' datasets that contain enough personal information can result in individuals being identifiable. Scholars and technologists continue to wrestle with how to adequately protect individuals when analysing such datasets [...] These are important considerations because they link to the fundamental ethical principle of minimizing harm".

(Association of Internet Researchers 2012, 6–7)

We can find two important points here:

a) In the online world, the boundary between the private and the public realm is messy. The question therefore arises if all Twitter content can be considered public content, as in a newspaper, or if there may also be content that is more private and intended for a limited audience.

b) Anonymisation becomes difficult online because data is stored on servers and is searchable. In the case of Twitter, search engines such as backtweets (http://backtweets.com/) allow us to search for archived tweets. The anonymity of cited content therefore becomes difficult to ascertain.

But does this mean that any qualitative analysis and quoting from Twitter violates research ethics? Or does one have to attain informed consent for each tweet one uses from others? The AoIR-document points out the complexity of online research ethics, but it does not provide any guidelines on how to actually deal with such questions.

The British Sociological Association (BSA 2002) recommends in its *Statement of Ethical Practice* that researchers studying the Internet should keep themselves updated on relevant issues:

"Members should take special care when carrying out research via the Internet. Ethical standards for internet research are not well developed as yet. Eliciting informed consent, negotiating access agreements, assessing the boundaries between the public and the private, and ensuring the security of data transmissions are all problematic in Internet research. Members who carry out research online should ensure that they are familiar with ongoing debates on the ethics

Established Research Ethics Guidelines

of Internet research, and might wish to consider erring on the side of caution in making judgements affecting the well-being of online research participants".

(BSA 2002, §41)

This short paragraph certainly does not help an Internet researcher in any particular situation in which s/he deals with ethical issues. The International Sociological Association's 2001 *Code of Ethics* argues in respect to informed consent:

"The security, anonymity and privacy of research subjects and informants should be respected rigourously, in both quantitative and qualitative research. The sources of personal information obtained by researchers should be kept confidential, unless the informants have asked or agreed to be cited. Should informants be easily identifiable, researchers should remind them explicitly of the consequences that may follow from the publication of the research data and outcomes. [...] The consent of research subjects and informants should be obtained in advance".

The ISA code does not mention the specificities of online research. Anonymity often does not exist online. Obtaining informed content when working with a large online dataset is for the most part practically impossible due to time restrictions. In the online world, the private and the public spheres do not uphold clear boundaries.

The American Sociological Association's (ASA's) 1999 *Code of Ethics* says the following about anonymity and informed consent:

11.06 Anonymity of Sources

a) Sociologists do not disclose in their writings, lectures, or other public media confidential, personally identifiable information concerning their research participants, students, individual or organisational clients, or other recipients of their service which is obtained during the course of their work, unless consent from individuals or their legal representatives has been obtained.

b) When confidential information is used in scientific and professional presentations, sociologists disguise the identity of research participants, students, individual or organisational clients, or other recipients of their service.

[...]

12.01 Scope of Informed Consent

"(a) Sociologists conducting research obtain consent from research partici-pants or their legally authorized representatives (1) when data are collected from research participants through any form of communication, interaction, or intervention; or (2) when behavior of research participants occurs in a private context where an individual can reasonably expect that no observation or re-porting is taking place. [...] (c) Sociologists may conduct research in public places or use publicly-available information about individuals (e.g., naturalistic observations in public places, analysis of public records, or archival research) without obtaining consent".

(ASA 1999)

The ASA code does not specifically mention online research. It does not recognise that in online research it is not straightforward to keep cited content anonymous. However, it does make a good point in remarking that there is a difference in obtaining informed consent in respect to the question of whether communication, interaction, and behaviour take place in a private context or in a public place. In relation to social media, this means that one needs to ask which communications are private and which ones are public.

Overall, the discussion shows that established ethics guidelines do not direct much at-tention to the particularities of online research ethics.

5.3 Towards Critical-Realist Internet Research Ethics

There are two extremes in Internet research ethics. The one extreme argues that one must obtain informed consent for every piece of data one gathers online. The other ar-gues that what is online is out there and can and should be analysed without regard to ethical considerations.

Michael Zimmer (2010) discusses the question of whether or not it is ethical to harvest Twitter data without informed consent:

"Yes, setting one's Twitter stream to public does mean that anyone can search for you, follow you, and view your activity. However, there is a reasonable expectation that one's tweet stream will be "practically obscure" within the thousands (if not millions) of tweets similarly publicly viewable. Yes, the subject has consented to making her tweets visible to those who take the time and energy to seek her out, those who have a genuine interest to connect and view her activity through this social network".

"But she did *not* automatically consent, I argue, to having her tweet stream systematically followed, harvested, archived, and mined by researchers (no matter the positive intent of such research). That is not what is expected when making a Twitter account public, and it is my opinion that researchers should seek consent prior to capturing and using this data".

Some of the people commenting on this blog post heavily disagreed with Zimmer's perspective:

"It's like a blog. (Originally, Twitter was called 'the microblogging service'.) You can quote and attribute from blogs, but you can't pretend it's your work [...] As for someone deciding to analyse me from my tweets and publish the results – well, not much i can do about the analysis"

"The web is not an environment that supports a reasonable expectation of privacy in public. Unmistakeably not. Nor does twitter as a subculture gesture toward such an expectation".

"*Once tweeted, a birdsong is gone forever. No deleting or taking back what's been broadcast to the world. If someone seeks privacy, they should seek another method of communication*".

"TWITTER IS PUBLIC – NO QUESTION ABOUT IT. Tweets (from the public stream) are like to be treated like blogs (microblogs) and webpages – PUBLIC. No consent required for analyzing them, unless of course they are DMs (which are like emails – confidential) or sent to your "followers only". [...] You tweet because you want to get your message out, and not only to our friends (ever heard of retweets?)".

"This is VERY different from discussion boards, chat rooms, or even Facebook. [...] I simply dispute that ANYBODY who tweets (regardless of whether he has read the privacy policy or not) does so under the expectation of privacy or having a "limited" audience (if they want to do that, there is a privacy setting for that). Anybody who tweets sees on a daily basis that others are retweeting their tweets or quoting from their tweets also appear in search engines and on the twitter homepage itself".

The discussion shows that there is a conflict between research ethics fundamentalists and big data positivist. Research ethics fundamentalists tend to say:

"You have to attain informed consent for every piece of social media data you gather because we cannot assume automatic consent. Users tend not to read a platform's privacy policies – they may assume that some of their data is private, and they may not agree to their data being used in research. Even if you anonymise the users you quote, many might still be identified in the networked online environment".

There are limits of informed consent. Informed consent can censor critical research and cause harm for a researcher conducting critical online research if s/he contacts a user, asking: "Dear Mr. Misogynist/Nazi/Right-Wing Extremist etc.! I am a social researcher gathering data from Twitter. Can you please give me your informed consent for quoting your violent threat against X?". The researcher may be next in line for being harassed or threatened.

A solution would be to only use aggregated data. But such an approach is biased towards quantitative methods and computational social science. Critical discourse analysis and critical interpretative research thereby become impossible.

Big data positivists tend to say: "Most social media data is public data. It is like data in a newspaper. I can therefore gather big data without limits. Those talking about privacy want to limit the progress of social science". This position disregards any engagement with ethics and is biased towards quantification (meaning big data positivism, digital positivism). Zimmer and Proferes (2014) conducted a meta-study of 382 works focusing on Twitter research. Only 4% of the works discussed any ethical aspects. While privacy fetishism is one extreme, another its opposite pole is the complete ignorance of research ethics.

Privacy fetishism holds the danger of censoring and disabling critical research. It can endanger the critical researcher and result in violence directed against him/her by fascists, racists, nationalists, etc. Downright ignoring research ethics is often associated with a positivist approach to online research that focuses on the digital Lasswell formula: Who says what online, who do they say it to, how many likes, followers, re-tweets, comments, and friends do they have? The problem of this formula is that it leaves out questions such as the following: how are meanings expressed? What power structures condition the communication? What are the communicator's motivations, interests, and experiences? What contradictions does the communication involve?

We need critical-realist digital media research guidelines that go beyond research ethics fundamentalism and big data positivism. The approach needs to be realist in the sense that it avoids the two extremes of fundamentalism and positivism. The approach has to both engage with research ethics and enable the conduction of actual online research. The approach is critical in that it takes care to formulate guidelines in such a way as to enable and foster critical online research. By critical online research, we can understand any study that investigates digital media in the context of power structures (Fuchs 2017b).

In February 2016, I was part of a group of 16 scholars that met for a workshop funded by the Economic and Social Research Council (ESRC) at the University of Aberdeen. The task was that we create social media research ethics guidelines. The group consisted of a diverse range of scholars taking different perspectives on research ethics. Overall, the group managed to formulate some guidelines for a critical-realist research ethics framework (Townsend et al. 2016).

As one of the starting points for a realist perspective, we found a recommendation in the British Psychological Society's 2009 *Code of Ethics and Conduct* helpful: "Unless informed consent has been obtained, restrict research based upon observations of public behaviour to those situations in which persons being studied would reasonably expect to be observed by strangers" (BPS 2009, 13). The British Psychological Society's 2013 *Ethics Guidelines for Internet-Mediated Research* applies this principle to online research:

> "Where it is reasonable to argue that there is likely no perception and/or expectation of privacy (or where scientific/social value and/or research validity considerations are deemed to justify undisclosed observation), use of research data without gaining valid consent may be justifiable".
>
> (BPS 2013, 7)

Based on this insight, we formulated the following general guideline in the framework *Social Media Research: A Guide to Ethics*:

> "The question as to whether to consider social media data as private or public comes down, to some extent, to whether or not the social media user can reasonably expect to be observed by strangers (British Psychological Society, 2013; Fuchs, forthcoming). Things to consider here are: is the data you wish to

access on an open forum or platform (such as on Twitter), or is it located within a closed or private group (e.g. within Facebook) or a closed discussion forum? Is the group or forum password protected? Would platform users expect other visitors to have similar interests or issues to themselves? Does the group have a gatekeeper (or admin) that you could turn to for approval and advice? How have users set up their security settings? Data accessed from open and public online locations such as Twitter present less ethical issues than data which are found in closed or private online spaces. Similarly, data posted by public figures such as politicians, musicians and sportspeople on their public social media pages is less likely to be problematic because this data is intended to reach as wide an audience as possible. If the data you wish to access is held within a group for which you would need to gain membership approval, or if the group is password protected, there are more ethical issues to take into consideration".

(Townsend et al. 2016, 10)

Practically speaking, this means that analysing private messages and conversations in a closed group of recipients on Twitter requires informed consent. Most tweets, especially those using hashtags, aim at public visibility and therefore do not require informed consent in online research. How should one deal with Twitter users' identifiability? As good practice, one should not mention usernames, except for well-known public persons and institutions. One can instead use a pseudonym. It may still be possible to identify who posted a particular text that the researcher uses, but as this requires additional effort on the part of the person who wants to find out, the researcher does not directly identify the user.

Here is a specific example of how to apply these guidelines:

"Context: A researcher conducts a critical discourse analysis of a dataset of tweets using the hashtags #DonaldTrump; #TrumpTrain; #VoteTrump2016; #AlwaysTrump; #MakeAmericaGreatAgain or #Trum2016. These are analysed in order to find out how Trump supporters argue for their candidate on Twitter. Concerns: Can we consider this data public? Are there any issues of sensitivity or risk of harm? Do we need to seek informed consent before quoting these tweets directly?"

"Solution: Trump supporters use these hashtags in order to reach a broad public and convince other people to vote for Trump. It is therefore reasonable to assume that such tweets have public character: the authors expect and want to be observed by strangers in order to make a political point that they want others to read. The researcher can therefore directly quote such tweets without having to obtain informed consent. It is, however, good practice to delete the user IDs of everyday users, who are not themselves public figures".

(Townsend et al. 2016, 15)

5.4 Example Cases of Critical-Realist Internet Research Ethics

I want to outline an example of how I have dealt with research ethics in qualitative online studies that used critical discourse analysis. I will here deliberately abstract from the actual research results and merely focus on the ethical questions.

The study *Fascism 2.0: Twitter Users' Social Media Memories of Hitler on his 127th Birthday* (Fuchs 2017a) analysed how Twitter users communicated about Hitler on his 127th birthday. It utilised empirical ideology critique as its method. I used the tool Texifier to obtain all tweets from 20 April 2016 that mentioned any of the following hashtags: #hitler OR #adolfhitler OR #hitlerday OR #1488 OR #AdolfHitlerDay OR #HeilHitler OR #SiegHeil OR #HappyBirthdayAdolf OR #HitlerNation OR #HappyBirthdayHitler OR #HitlersBirthday OR #MakeGermanyGreatAgain OR #WeMissYouHitler. The search resulted in 4,193 tweets that were automatically imported into Discovertext, from where I exported them along with meta-data into a csv file. Using such hashtags on Hitler's birthday clearly aims at creating public attention. We can therefore say that the use of these hashtags in the context of Hitler's birthday constitutes a public space. Informed consent for analysing such postings is therefore not needed.

The study *Red Scare 2.0: User-Generated Ideology in the Age of Jeremy Corbyn and Social Media* (Fuchs 2016b) asked: how has Jeremy Corbyn during the Labour Leadership Election been framed in an ideological manner in discourses on Twitter and how have such ideological discourses been challenged? The study stands in the context of the negative framing of Corbyn during and following his run for the Labour Party leadership. With the help of Discovertext, I gathered 32,298 tweets based on the following search query: Corbyn AND anti-Semite OR anti-Semitic OR chaos OR clown OR commy OR communism OR communist OR loony OR Marx OR Marxist OR pinko OR red OR reds OR socialism OR socialist OR Stalin OR Stalinist OR terrorist OR violent OR violence. The data gathering was active for 23 days, from 22 August 2015 (23:25 BST) to 13 September 2015 (12:35

TABLE 5.1 The most active and most mentioned users in the Corbyn dataset

Users with largest no. of tweets	Frequency	Most mentioned users	Frequency
Redscarebot	322	anonymous2 (UKIP supporter)	723
Mywoodthorpe	241	ggreenwald	689
Ncolewilliams	237	independent	552
Houseoftwits	51	davidschneider	324
houseoftwitscon	43	rupertmurdoch	323
Gcinews	38	jeremycorbyn	311
anotao_news, anotao_nouvelle	37	telegraph	284
sunnyherring1	34	RT_com	221
anonymous1 (Corbyn-supporter)	32	edsbrown	215
Friedrichhayek	32	uklabour	212

BST). Corbyn was announced as the winner on 12 September 2015 (11:45 BST). It is reasonable to assume that users who tweet about Jeremy Corbyn during times when he is subject to increased public attention are directing their communication at the public. Also in this case, informed consent is therefore not required.

The study entailed a focus on the ten most active and most mentioned pro- and anti-Corbyn users (see Table 5.1). In the analysis, I anonymised individual users who are not well-known public figures and did not anonymise public figures (such as Glenn Greenwald, Rupert Murdoch, and David Schneider) and institutions (such as the Daily Telegraph, Russia Today, and The Independent).

The most active users were Twitter bots (redscarebot, mywoodthorpe). A bot based on an algorithm conducts certain online behaviour. Given that technologies do not maintain ethics, they likewise do not have expectations about privacy. They therefore do not need to be anonymised.

The study *Racism, Nationalism and Right-Wing Extremism Online: The Austrian Presidential Election 2016 on Facebook* (Fuchs 2016a) stands in the context of the Austrian presidential election 2016 that saw a run-off between the Green party candidate Alexander Van der Bellen and the Freedom Party of Austria's (FPÖ) far-right candidate Norbert Hofer. The paper asks: how did voters of Hofer express their support on Facebook? The FPÖ is the prototype of a European far-right party that bases its ideology on nationalism and xenophobia. Under the leadership of Jörg Haider (1986–2000), it was expanding and growing in popularity. Its current leader is Heinz Christian Strache.

Example Cases of Critical-Realist Internet Research Ethics

I used Netvizz in order to collect comments on postings related to Hofer's presidential candidacy. I accessed Norbert Hofer and Heinz Christian Strache's Facebook pages on 30 May 2016, and used Netvizz for extracting comments to postings made between 25 and 30 May. Given that the collected comments were posted in the days after the presidential election's second round, it is likely that the dataset contains data that refers to the political differences between Hofer and Van der Bellen. I selected postings by Hofer and Strache that were particularly polarising. This selection resulted in a total of 15 postings: 10 by Strache, 5 by Hofer. There were a total of 6,755 comments posted as responses to these 15 Facebook postings. So the analysed dataset consisted of 6,755 items.

The Facebook pages of Norbert Hofer and Heinz Christian Strache are public pages. All postings and comments on these pages are visible to everyone visiting them, not just to those who "like" them. One does not have to have a Facebook profile to access the two pages, as they can also be viewed without logging into Facebook. All postings and comments are thus visible in public. Furthermore, politicians are public figures. Citizens expect them to be present in the public. This includes that they post in public on social media and offer possibilities for public communication on their profiles. Given the public character of Strache and Hofer's Facebook pages, it is reasonable to assume that someone posting a comment on such a page can expect to be observed by strangers. In such a case, a researcher does not have to obtain informed consent for analysing and quoting comments. Given that the users are not public figures themselves, but only make public comments when posting on a politician's public Facebook page, I do not mention the usernames in the analysis. Netvizz does not save the usernames and so the collected dataset does not contain any identifiers.

5.5 Conclusion

Objectively speaking, the far right is fairly effective when it comes to utilising social media for political communication. Yet if one looks at the body of works published in social movement media studies, one gets the impression that political communication in the Internet age is by far dominated by politically progressive, left-wing, and social movements. There are comparatively few studies that focus on the Internet and far-right politics (Caiani and Kröll 2015). The far-right's use of the Internet has hardly been studied and is a blind spot in social movement media studies. W. Lance Bennett and Alexandra Segerberg's book The *Logic of Connective Action* (2013) mentions Occupy 70 times, but the Golden Dawn, Jobbik, the National Front, UKIP, Svoboda, Farage, or Le Pen not a single time. The *Encyclopedia of Social Movement Media* (Downing 2011) presents 600 pages analyses of "alternative media, citizens' media, community media, counterinformation media, grassroots media, independent media, nano-media, participatory media, social movement media, and underground media" (Downing 2011, xxv). The focus here is on all sorts of progressive and left-wing media, from the likes of the Adbusters Media Foundation to Zapatista media. The editor John Downing (2011, xxvi) admits that "much less examination of media of extreme right movements occurs in this volume than there might be", but he does not explain why this is the case, why it is problematic, and how it could be changed.

Most social movement researchers like to do feel-good research. They study progressive left-wing movements that they like and are sympathetic towards, consider such studies as a form of solidarity, and tend to simply celebrate how these groups organise and communicate. Such studies make the researchers feel good and politically engaged. But celebratory studies of these movements hardly help us to understand the difficult contradictions that left-wing activism faces in the capitalist world. They neglect analysing right-wing movements and groups that pose a threat to democracy. And thus this is the blind spot of social movement media studies.

One might now be tempted to argue that far-right groups are not part of social movement studies because they tend to be hierarchic, have a populist leader, and aim at a society that is governed from the top in an authoritarian or even fascist manner. However, such a definitional exclusion overlooks that also left-wing progressive movements often develop certain hierarchies and forms of leadership. Left-wing movements too attempt to define the social as a progressive political concept by arguing that the far right has anti-social political goals. The "social" in social movements means nothing more than the circumstance that social movements are groups that act collectively in order to change society and move it in a certain direction. It tells us nothing about these groups' political content. The point is that in a contradictory world, social movements are contradictory. They contest how society is developing. Two options that are today possible are the democratic socialist option of participatory democracy and the authoritarian option of fascist barbarism. Social movement studies should focus on studying the diverse range of political movements.

Conclusion

Studying online politics poses ethical challenges in respect to privacy/the public, anonymity, and informed consent. Conventional research ethics guidelines often ignore qualitative online research or have little to say on the topic. Conducting studies of online nationalism, racism, xenophobia, and fascism poses additional challenges because these phenomena are inherently violent. Debates on Internet research ethics face two extremes. On the one side, research ethics fundamentalism obstructs qualitative online research. On the other, big data positivism lacks a critical focus on qualitative dimensions of analysis. The alternative is a critical-realist online research ethics that informs critical studies of digital media.

References

American Sociological Association. 1999. Code of Ethics and Policies and Procedures of the ASA Committee on Professional Ethics. http://www.asanet.org/sites/default/files/code_of_ethics.pdf

Association of Internet Researchers. 2012. Ethical Decision-Making and Internet Research. http://aoir.org/reports/ethics2.pdf

Bennett, W. Lance and Alexandra Segerberg. 2013. *The Logic of Connective action. Digital Media and the Personalization of Contentious Politics.* Cambridge: Cambridge University Press.

British Psychological Society. 2013. Ethics Guidelines for Internet-Mediated Research. http://www.bps.org.uk/system/files/Public%20files/inf206-guidelines-for-internet-mediated-research.pdf

British Psychological Society. 2009. Code of Ethics and Conduct. http://www.bps.org.uk/system/files/documents/code_of_ethics_and_conduct.pdf

British Sociological Association. 2002. Statement of Ethical Practice. https://www.britsoc.co.uk/equality-diversity/statement-of-ethical-practice

Caiani, Manuela and Patricia Kröll. 2015. The Transnationalization of the Extreme Right and the Use of the Internet. *International Journal of Comparative and Applied Criminal Justice* 39 (4): 331–351.

Downing, John D. H., ed. 2011. *Encyclopedia of Social Movement Media.* Thousand Oaks, CA: Sage.

Fuchs, Christian. 2017a. Fascism 2.0: Twitter Jsers' Social Media Memories of Hitler on his 127th Birthday. *Fascism: Journal of Comparative Facist Studies* 6 (2): 228–263.

Fuchs, Christian. 2017b. *Social Media: A Critical Introduction.* London: Sage. Second edition.

Fuchs, Christian. 2016a. Racism, Nationalism and Right-Wing Extremism Online: The Austrian Presidential Election 2016 on Facebook. *Momentum Quarterly – Zeitschrift für sozialen Fortschritt (Journal for Societal Progress)* 5 (3): 172–196.

Fuchs, Christian. 2016b. Red Scare 2.0: User-generated Ideology in the Age of Jeremy Corbyn and Social Media. *Journal of Language and Politics* 15 (4): 369–398.

International Sociological Association. 2001. Code of Ethics. http://isa-sociology.org/en/about-isa/code-of-ethics/

Townsend, Leanne et al. 2016. *Social Media Research: A Guide to Ethics.* http://www.gla.ac.uk/media/media_487729_en.pdf

Zimmer, Michael. 2010. Is It Ethical to Harvest Public Twitter Accounts without Consent? http://www.michaelzimmer.org/2010/02/12/is-it-ethical-to-harvest-public-twitter-accounts-without-consent/

Zimmer, Michael and Nicholas John Proferes. 2014. A Topology of Twitter Research: Disciplines, Methods, and eThics. *Aslib Journal of Information Management* 66 (3): 250–261.

Chapter Six
Towards an Alternative Concept of Privacy

6.1 Introduction

"How Privacy Vanishes Online"

(*New York Times*; March 16, 2010)

"Critics Say Google Invades Privacy with New Service"

(*New York Times*; February 12, 2010)

"Privacy No Longer A Social Norm, Says Facebook Founder"

(*The Guardian*; January 11, 2010)

"Google's Eric Schmidt: Privacy is Paramount"

(*The Guardian*; March 19, 2010)

"Google Street View: Politicians Insist on Privacy"

(*Süddeutsche Zeitung*; August 12, 2010)

These example news clippings show that privacy is a much-talked-about issue in the contemporary information society. The problem of discussions about privacy in the media and the public is frequent that a clear understanding of privacy is missing. Privacy tends to be conceived as a universal and always positive value. The downsides of privacy tend to be neglected. The task of this chapter is to show some limits of the privacy concept by giving a critical political economy analysis of this notion. Such an approach is especially interested in uncovering the role of surplus value, exploitation, and class in the studied phenomena (Dussel 2008, 77; Negri 1991, 74).

DOI: 10.4324/9781003279488-8

One task of this chapter is to provide some meta-theoretical reflections about how a privacy typology can best be grounded (Section 6.2). The Marxian notion of fetishism is employed for challenging naturalisations of privacy (Section 6.3). A more systematic critique of the modern privacy concept is elaborated (Section 6.4). The Critical Political Economy approach reminds us that it is important not to see only the positive aspects of privacy, but also its downsides. But such an analysis brings up the question if one should drop the privacy concept or if it is possible to establish an alternative privacy concept that avoids the current limits. This question is outlined in the concluding section of this work (Section 6.5). The major novelty of this chapter is that it uses Critical Political Economy for working out the theoretical foundations of a socialist privacy concept.

6.2 The Privacy Concept

Jürgen Habermas (1987, 318–321) has provided a model of modern society that allows conceptualising the relation of the private and the public. The lifeworld is made up of the private sphere (family, private households, intimacy) and the public sphere (communicative networks that enable private persons to take part in culture and the formation of public opinion; Habermas 1987, 319). The lifeworld is coupled in various ways to the economic system and the administrative system that are steered by money and power. In modern society, the economy has become disembedded from the private sphere and has formed a separate subsystem of society (Habermas 1989). The private sphere is for Habermas the sphere, where the individual is most in control of his activities and communications. Discussions about privacy typically relate to the question if and to which extents humans are able to control data that is generated about them. It shares the value of private self-determination characteristic for the private sphere. The distinction between the private and the public spheres is relevant for the privacy concept because it entails "the distinction between things that should be shown and things that should be hidden" (Arendt 1958, 72).

Ferdinand Schoeman (1984b, 2f) distinguishes between definitions of privacy as (1) claim, entitlement, right, (2) the measure of control of an individual over personal information, intimacy, and visibility, (3) state or condition of limited access to an individual. Ken Gormley (1992, 1337f) discerns four types of privacy definitions: (1) privacy as an expression of one's personhood and personality, (2) privacy as autonomy, (3) privacy as citizens' ability to regulate information about themselves, (4) multidimensional notions of privacy. Daniel Solove (2008) says that there are six different privacy definitions: privacy as (1) the right to be left alone, (2) limited access to the self, (3) secrecy, (4) control over personal information, (5) personhood, and (6) intimacy. Solove (2004) distinguishes between the following conceptions: (1) privacy as protection from Big Brother, (2) privacy as secrecy, (3) privacy as non-invasion, (4) privacy as control over information use. The problem with these privacy typologies is that they are arbitrary: there is no theoretical criterion used for distinguishing the differences between the categories. The different definitions are postulated, but not theoretically grounded. A theoretical criterion is missing that is used for distinguishing different ways of defining privacy. Providing such an analysis is a meta-theoretical task.

Anthony Giddens sees the "division between objectivism and subjectivism" (Giddens 1984, xx) as one of the central issues of social theory. Subjective approaches are oriented on human agents and their practices as a primary object of analysis, objective approaches on social structures. Structures in this respect are institutionalised relationships that are stabilised across time and space (Giddens 1984, xxxi).

Herman Tavani (2008) distinguishes between restricted access theories, control theories, and restricted access/limited control theories of privacy. The restricted access theory of privacy sees privacy given if one is able to limit and restrict others from access to personal information and personal affairs (Tavani 2008, 142ff). The classical form of this definition is Warren's and Brandeis' notion of privacy: "Now the right to life has come to mean the right to enjoy life, – the right to be let alone" (Warren and Brandeis 1890, 193). They discussed this right especially in relation to newspapers and spoke of the "evil of invasion of privacy by the newspapers". Although some scholars argue that Warren's and Brandeis' (1890) paper is the source of the restricted access theory (for example, Bloustein 1964/1984; Rule 2007, 22; Schoeman 1984b; Solove 2008, 15f), the same concept was already formulated by John Stuart Mills 42 years before Warren and Brandeis in his 1848 book *Principles of political economy* (Mill 1965, 938):

> "That there is, or ought to be, some space in human existence thus entrenched 'around', and sacred from authoritative intrusion, no one who professes the smallest regard to human dignity will call in question: the point to be determined is, where the limit should be placed; how large a province of human life this reserved territory should include".

(Mill 1965, 938)

This circumstance shows the inherent connection between the modern privacy concept and liberal thought. Restricted access definitions of privacy can for example be found in the works of Allen (1988, 3), Bok (1983, 10), Gavison (1980, 428f), Nock (1993, 1), and Schoeman (1992, 15, 106, 107f).

The Privacy Concept

Restricted access theories conceive privacy as transsubjective, it is seen as an objective normative right or moral value that exists also if politics or human practices choose to implement mechanisms that reveal private facts in public or allow the public access to the private sphere of individuals. Privacy is considered as a moral structure that is aimed at protecting all humans. Therefore Mill conceives privacy as a circle around individuals:

> "There is a circle around every individual human being, which no government, be it that of one, of a few, or of the many, ought to be permitted to overstep: there is a part of the life of every person who has come to years of discretion, within which the individuality of that person ought to reign uncontrolled either by any other individual or by the public collectively".
>
> (Mill 1965, 938)

The restricted access theory of privacy is an objective theory in Giddens' terminology because it conceives privacy as a transindividual moral structure that exists as a right and ethical imperative relatively independently of single human actions.

The control theory of privacy sees privacy as control and self-determination over information about oneself and over the access to one's personal affairs (Tavani 2008, 142ff). The most well-known privacy theory of this kind was formulated by Alan Westin, who defined privacy as the "claim of individuals, groups or institutions to determine for themselves when, how, and to what extent information about them is communicated to others" (Westin 1967, 7). Other examples are the definitions given by Fried (1968/1984, 209), Froomkin (2000, 1464), Miller (1971, 25), Quinn (2006, 214), Schultz (2006, 108), Rule (2007, 3), Shils (1966, 281f), Solove (2004, 51), and Spinello (2006, 143). Control theories are focused on individual self-determination over privacy. Privacy is dependent on human action, individuals may choose to withhold or reveal a lot of information about themselves. Privacy in these theories is therefore variable, dynamic, and flexible, depending on the behaviour of individuals. Control theories of privacy are subjective theories in Giddens' terminology because they stress the dependence of privacy on human subjectivity and individual action and choosing.

The restricted access/limited control theory (RALC) of privacy tries to combine both concepts. It distinguishes "between the concept of privacy, which it defines in terms of restricted access, and the management of privacy, which is achieved via a system of limited controls for individuals" (Tavani 2008, 144). James H. Moor, on the one hand, uses privacy to "designate a situation in which people are protected from intrusion or observation", and, on the other hand, speaks of "different zones of privacy", in which "one can decide how much personal information to keep private and how much to make public. [...] Different people may be given different levels of access for different kinds of information at different times" (Moor 2000, 207f; see also: Introna 2000, 190; Shade 2003, 278).

Giddens (1984) has tried to overcome the separation of subject and object in his theory of structuration by formulating the theorem of the duality of structure that connects subjects and objects of society dialectically by arguing that social structures are medium and outcome of social actions, at the same time they enable and constrain practices (Fuchs 2003a, 2003b). Applying this theorem to privacy gives a good description of the RALC: control refers to the human agency level of privacy that enables the existence of a protective sphere, which enables humans to act in society with a degree of protection into their private affairs. Limited access refers to a moral structural sphere that protects individuals from privacy intrusion and enables them to act in society. The RALC sees restricted access and individual control as mutually constitutive. Individuals and society may choose to regulate privacy in certain ways, which is an aspect of subjectivity and action. Based on this action, a sphere of privacy of individuals that is protected from access to others may be set up that enables individuals to act in society, their private sphere, and the public based on privacy and data protection. Privacy control as human action may under certain circumstances or in given contexts (Nissenbaum 2010; Solove 2008[1]) then change the degree of access (Table 6.1).

TABLE 6.1 A typology of privacy theories

Theoretical criterion	Approach	Description
Subjectivism	Control	Privacy as individual control and self-determination of the access of others to one's private sphere.
Objectivism	Restricted access	Privacy as the right or norm of restricting others' access to one's personal affairs.
Subject/object-dialectic	Restricted access/control	Privacy as process, in which action regulates and manages the conditions of the private sphere and can thereby enable the existence of a protective sphere that allows individuals to act in society.

The Privacy Concept

For a critical analysis, it does not suffice to understand, which ways there are of defining privacy. It is also necessary to discuss the limits and problems of the privacy approach. For doing so, I will introduce the notion of privacy fetishism in the next section.

6.3 Privacy Fetishism

Etzioni (1999) stresses that it is a typical American liberal belief that strengthening privacy can cause no harm. He stresses that privacy can undermine common goods (public safety, public health). That privacy is not automatically a positive value has also been reflected in criticism of privacy. Critics of the privacy concept argue that it promotes an individual agenda and possessive individualism that can harm the public/common good (Bennett 2008, 9f; Bennett and Raab 2006, 14; Etzioni 1999; Gilliom 2001, 8, 121; Hongladarom 2007, 115; Lyon 1994, 296; 2007, 7, 174; Stalder 2002; Tavani 2008, 157f), that it can be used for legitimatising domestic violence in families (Bennett and Raab 2006, 15; Lyon 1994, 16; 2001, 20; Quinn 2006, 214; Schoeman 1992, 13f; Tavani 2008, 157f; Wacks 2010, 36), that it can be used for planning and carrying out illegal or antisocial activities (Quinn 2006, 214; Schoeman 1984b, 8), that it can conceal information in order to mislead and misrepresent the character of individuals and is therefore deceptive (Bennett and Raab 2006, 14; Schoeman 1984a, 403; Posner 1978/1984; Wasserstrom 1978/1984), that a separation of public and private life is problematic (Bennett and Raab 2006, 15; Lyon 2001, 20; Sewell and Barker 2007, 354f), and that it advances a liberal notion of democracy that can be opposed by a participatory notion of democracy (Bennett and Raab 2006, 15). Privacy has also been criticised as a Western-centric concept that does not exist in an individualistic form in non-Western societies (Burk 2007; Hongladarom 2007; Zureik and Harling Stalker 2010, 12). There have also been discussions of the concept of privacy based on ideology critique (Stahl 2007) and intercultural philosophy (see for example, Capurro 2005; Ess 2005).

These critiques show that the question is therefore not how privacy can be best protected, but in which cases whose privacy should be protected and in which cases it should not be protected. Many constitutional privacy regulations acknowledge the limits of privacy and private property and that unlimited property can harm the public good. So the fifth amendment of the US constitution says that no person shall "be deprived of life, liberty, or property", but adds: "without due process of law". It says that private property shall not "be taken for public use, without just compensation". Article 14 (1) of the German Grundgesetz says that the "property and the right of inheritance shall be guaranteed" and adds: "(2) Property entails obligations. Its use shall also serve the public good. (3) Expropriation shall only be permissible for the public good". Similarly, the

Swedish Constitution (The Instrument of Government, Chapter 2) guarantees "the property of every citizen", but adds that this is not the case when expropriation is "necessary to satisfy pressing public interests" (§18).

Liberal privacy theories typically talk about the positive qualities that privacy entails for humans or speak of it as an anthropological constant in all societies, without discussing the particular role of privacy in modern, capitalist society. Alan Westin (1967) on the one hand gives examples from anthropological literature of societies without privacy, but on the other hand in contradiction to his own examples claims that privacy is a universal phenomenon that can be found in sexual relations, households, personal encounter, religion, puberty, and that is related to gossip and curiosity.

Bloustein (1964/1984) argues that privacy is needed for protecting individual dignity, integrity, independence, freedom, and self-determination. For Westin (1967), privacy provides individual autonomy, emotional release, self-evaluation, and intimacy. Fried (1968/1984) sees privacy as a context that enables human respect, love, friendship, and trust. Benn (1971/1984) says that privacy is a general principle needed for respect, freedom, and autonomy. For Rachel (1975/1984), privacy is needed for protecting individuals from competition and embarrassment. Gerstein (1978/1984) argues that intimacy cannot exist without privacy. For Gavison (1980), privacy protects freedom from physical access, liberty of action, freedom from censure and ridicule, and promotes mental health, autonomy, human relations, dignity, pluralism, tolerance, and democracy. Ferdinand Schoeman (1984a) argues that privacy enables social relationships, intimacy, personality, and personally validated objectives that are autonomously defined. Margulis (2003a, 2003b) says that privacy enables autonomy, emotional release, self-evaluation, and protected communication. Solove (2008, 98) argues that privacy is a pluralistic value and provides a list of the values privacy has been associated with: autonomy, counterculture, creativity, democracy, eccentricity, dignity, freedom, freedom of thought, friendship, human relationships, imagination, independence, individuality, intimacy, psychological well-being, reputation, self-development. Given the preceding discussion, the following values can be added to this list: emotional release, Individual integrity, love, personality, pluralism, self-determination, respect, tolerance, self-evaluation, and trust.

Such analyses do not engage with actual and possible negative effects of the privacy and the relationship of modern privacy to private property, capital accumulation, and social inequality. They give unhistorical accounts of privacy by arguing that privacy is a universal human principle that brings about positive qualities for individuals and society. They abstract from issues relating to the political economy of capitalism, such as exploitation and income/wealth inequality. But if there are negative aspects of modern privacy, such as the shielding of income gaps and of corporate crimes, then universalistic liberal privacy accounts are problematic because they neglect negative aspects and present modern values as characteristic for all societies. Karl Marx characterised the appearance of the "definite social relation between men themselves" as "the fantastic form of a relation between things" (Marx 1867, 167) as fetishistic thinking. Fetishism mistakes phenomena that are created by humans and have social and historical character as being natural and existing always and forever in all societies. Phenomena such as the commodity are declared to be "everlasting truths" (Marx 1867, 175, fn34). Theories of privacy that do not consider privacy as historical, that do not take into account the relation of privacy and capitalism or only stress its positive role, can, based on Marx, be characterised as privacy fetishism. In contrast to privacy fetishism, Moore (1984) argues based on anthropological and historical analyses of privacy that it is not an anthropological need "like the need for air, sleep, or nourishment" (Moore 1984, 71), but "a socially created need" that varies historically (Moore 1984, 73). The desire for privacy, according to Moore, develops only in societies that have a public sphere that is characterised by complex social relationships that are seen as "disagreeable or threatening obligation" (Moore 1984, 72). Moore argues that this situation is the result of stratified societies, in which there are winners and losers. The alternative would be the "direct participation in decisions affecting daily lives" (Moore 1984, 79).

This discussion reminds us that it is important to contextualise privacy. If it is indeed true that we live in a capitalist society, as the recent world economic crisis has again made clear, then it is important to analyse privacy in the context of the political economy of capitalism. The next section will cover this topic.

6.4 The Limits of Privacy in Capitalism

Privacy has been characterised as a value that is typical for liberal worldviews (Bennett and Raab 2006, 4, 17; Etzioni 1999; Moore 1984, 75f). It is therefore no surprise that John Stuart Mill has in his political economy introduced the notion of privacy in relation to private property. When discussing the conditions under which land should be allowed to be transformed into private property, he speaks of the necessity of "the owner's privacy against invasion" (Mill 1965, 232).

Karl Marx and Friedrich Engels worked out an early critique of liberal privacy concepts. This critique contains four central elements. The critique of privacy by Marx and Engels has not been covered in the literature in any detail. Therefore the outline of this critique is deliberately strongly quotation based in order to make their critique available in the form of a comprehensive overview.

1) There is no pure individual existence. All human existence is socially conditioned. By conceiving privacy as an individual right, liberal privacy conceptions fail to grasp the social existence of humans.

   Marx described the position of the relation of the private and the general in the theories of bourgeois political economists:

   > "The economists express this as follows: Each pursues his private interest and only his private interest; and thereby serves the private interests of all, the general interest, without willing or knowing it. [...] The point is rather that private interest is itself already a socially determined interest, which can be achieved only within the conditions laid down by society and with the means provided by Marx 1857/58, 156".

   Marx argues that the notion of the private in classical political economy is individualistic and neglects that all individual actions take place within and are conditioned by society.

2) The individualism advanced by liberal privacy theories results in egoism that harms the public good.

   Marx furthermore stresses that modern society is not only based on individualism, but also on egoism (Marx 1843b, 235–237, 240). Liberty in bourgeois society is "is the liberty of man viewed as an isolated monad, withdrawn into himself. [...] The practical application of the right of liberty is the right of private property" (Marx 1843b, 235). Modern society's constitution would be the "constitution of private property" (Marx 1843a, 166). The right of private property in the means of production and to accumulate as much capital as one pleases would harm the community and the social welfare of others who are by this process deprived of wealth: "The right of property is thus the right to enjoy and dispose one's possessions as one wills, without regard for other men and independently of society. It is the right of self-interest" (Marx 1843b, 236). "Thus none of the so-called rights of men goes beyond the egoistic man, the man withdrawn into himself, his private interest and his private choice, and separated from the community as a member of civil society" (Marx 1843b, 236f). Marx further criticises that the private accumulation of capital results in the concentration of capital and thereby of wealth:

The Limits of Privacy in Capitalism

"Accumulation, where private property prevails, is the concentration of capital in the hands of a few" (Marx 1844, 41).

David Lyon notes that the liberal

> "conception of privacy connects neatly with private property. Mill's sovereign individual were characterized by freedom to pursue their own interests without interference [...]. This presupposes a highly competitive environment, in which one person's freedom would impinge on another's, hence the need to balance values like 'privacy' with others".
>
> (Lyon 1994, 186)

Crawford Macpherson (1962) has termed this Marxian critique of liberalism the critique of possessive individualism. Possessive individualism is the "conception of the individual as essentially the proprietor of his own person or capacities, owing nothing to society for them" (Macpherson 1962, 3). According to Macpherson, it is the underlying worldview of liberal-democratic theory since John Locke and John Stuart Mill. The problem of the liberal notion of privacy and the private sphere is that relatively unhindered private accumulation of wealth, as the neoliberal regime of accumulation has shown since the 1970s, comes into conflict with social justice and is likely to result in strong socio-economic inequality. The ultimate result of Mill's understanding of privacy is an extremely unequal distribution of wealth. So his privacy concept privileges the rich owning class at the expense of the non-owners of private property in the means of production.

3) The concepts of privacy and the private sphere are ideological foundations of the modern class structure.

Marx says that capitalism's "principle of individualism" and a constitution of state and society that guarantees the existence of classes is the attempt "to plunge man back into the limitations of his private sphere" (Marx 1843a, 147) and to thereby make him a "private human being" (Marx 1843a, 148). If the private sphere in modern society is connected to the notion of private property, then it is an inherent foundation of the class antagonism between capital and work: "But labor, the subjective essence of private property as exclusion of property, and capital, objective labor as exclusion of labor, constitute private property as its developed state of contradiction-hence a dynamic relationship moving inexorably to its resolution" (Marx 1844, 99). The capitalist mode of production is on the one hand based on the "socialization of labour" and "socially exploited and therefore communal means of production" (Marx 1867, 928). This social dimension of capitalism is circumvented by private ownership of the means of production: "Private property, as the antithesis to social, collective property, exists only where the means of labour and the external conditions of labour belong to private individuals" (Marx 1867, 927). "But modern bourgeois private property is the final and most complete expression of the system of producing and appropriating products, that is based on class antagonisms, on the exploitation of the many by the few" (Marx and Engels 1846, 484).

4) There is an inherent connection of privacy, private property, and the patriarchal family.

Engels (1891, 474, 480) has stressed the inherent connection of the private sphere with private property and the patriarchal family.

The Marxian analysis of the political economy of privacy was partly reflected in the works of Jürgen Habermas and Hannah Arendt.

Marx stresses that capitalism is based on a separation of the state and bourgeois society. The latter would be based on private property. Man

"leads a double life. [...] In the political community he regards himself as communal being; but in civil society he is active as a private individual, treats other men as means, reduces himself to a means, and becomes the plaything of alien powers".

(Marx 1843b, 225; see also: Marx 1843a, 90)

This Marxian moment of analysis is a crucial element in Habermas' theory of the public sphere. During the course of the development of capitalism since the 19th century, the world of work and organisation became a distinct sphere. With the rise of wage labour, industrialism, and the factory, the economy became to a certain degree disembedded from the private household (Habermas 1989, 152, 154; see also: Arendt 1958, 47, 68). Consumption became a central role of the private sphere:

"On the other hand, the family now evolved even more into a consumer of income and leisure time, into the recipient of publicly guaranteed compensations and support services. Private autonomy was maintained not so much in functions of control as in functions of consumption".

(Habermas 1989, 156)

Therefore, privacy is for Habermas an illusionary ideology – "pseudo-privacy" (Habermas 1989, 157) – that in reality functions as a community of consumers: "there arose the illusion of an intensified privacy in an interior domain whose scope had shrunk to

The Limits of Privacy in Capitalism

compromise the conjugal family only insofar as it constituted a community of consumers" (Habermas 1989, 156). A central role of the private sphere in capitalism is also that it is a sphere of leisure: "Leisure behavior supplies the key to the floodlit privacy of the new sphere, to the externalisation of what is declared to be the inner life" (Habermas 1989, 159). Expressed in other words, one can say that the role of the private sphere in capitalism as the sphere of leisure and consumption that Habermas identifies is that it guarantees the reproduction of labour power so that it remains vital, productive, and exploitable. Habermas (1989, 124–129) stresses that for Marx the inherent principle of universal accessibility of the public sphere is undermined by the fact that in capitalism private property of the means of production is controlled by capitalists and workers are excluded from this ownership. The separation of the private from the public realm obstructs "what the idea of the bourgeois public sphere promised" (Habermas 1989, 125).

Hannah Arendt (1958) reflects in her work the Marxian notion that the liberal privacy concept is atomistic and alienates humans from their social essence. She stresses that sociality is a fundamental human condition. Privacy is for her in modern society "a sphere of intimacy" (Arendt 1958, 38). For Arendt, the public realm is a sphere, where everything can be seen and heard by everybody (Arendt 1958, 50). It is "the common world" that "gathers us together and yet prevents our falling over each other" (Arendt 1958, 52). Privacy would be a sphere of deprivation, where humans are deprived of social relations and "the possibility of achieving something more permanent than life itself" (Arendt 1958, 58). "The privation of privacy lies in the absence of others" (Arendt 1958, 58). Arendt says that the relation between private and public is "manifest in its most elementary level in the question of private property" (Arendt 1958, 61). In modern society, as a result of private property, the public would have become a function of the private and the private the only common concern left, a flight from the outer world into intimacy (Arendt 1958, 69). Labour and economic production, formerly part of private households, would have become public by being integrated into capitalist production.

The theories of Marx, Arendt, and Habermas have in common that they stress the importance of addressing the notions of privacy and the public by analyzing their inherent connection to the political economy of capitalism.

The connection between privacy and private property becomes apparent in countries like Switzerland, Liechtenstein, Monaco, or Austria that have a tradition of the relative anonymity of bank accounts and transactions. Money as private property is seen as an aspect of privacy, about which no or only restricted information should be known to the public. In Switzerland, the bank secret is defined in the Federal Banking Act (§47). The Swiss Bankers Association sees bank anonymity as a form of "financial privacy" (http://www.swissbanking.org/en/home/qa-090313.htm) that needs to be protected and of "privacy in relation to financial income and assets" (http://www.swissbanking.org/en/home/dossier-bankkundengeheimnis/dossier-bankkundengeheimnis-themen-geheimnis. htm). In most countries, information about income and the profits of companies (except for public companies) is treated as a secret, a form of financial privacy. The problems of secret bank accounts and transactions and the intransparency of richness and company are not only that secrecy can in the economy support tax evasion, black money, and money laundering, but that it masks wealth gaps. Financial privacy reflects the classical liberal account of privacy. So for example John Stuart Mill formulated a right of the propertied class to economic privacy as "the owner's privacy against invasion" (Mill 1965, 232). Economic privacy in capitalism (the right to keep information about income, profits, bank transactions secret) protects the rich, companies, and wealthy. The anonymity of wealth, high incomes, and profits makes income and wealth gaps between the rich and the poor invisible and thereby ideologically helps legitimatising and upholding these gaps. It can therefore be considered an ideological mechanism that helps reproducing and deepening inequality.

Privacy is in modern societies an ideal rooted in the Enlightenment. The rise of capitalism resulted in the idea that the private sphere should be separated from the public sphere and not accessible for the public and that therefore autonomy and anonymity of the individual is needed in the private sphere. The rise of the idea of privacy in modern society is connected to the rise of the central ideal of the freedom of private ownership. Private ownership is the idea that humans have the right to own as much wealth as they want, as long as it is inherited or acquired through individual achievements. There is an antagonism between private ownership and social equity in modern society. How much and what exactly a person owns is treated as an aspect of privacy in contemporary society. To keep ownership structures secret is a measure of precaution against the public questioning or the political and individual attack against private ownership. Capitalism requires anonymity and privacy in order to function. But full privacy is also not possible in modern society because strangers enter social relations that require trust or enable exchange. Building trust requires knowing certain data about other persons. It is therefore checked with the help of surveillance procedures if a stranger can be trusted. Corporations have the aim of accumulating ever more capital. That is why they have an interest in knowing as much as possible about their workers (in order to control

The Limits of Privacy in Capitalism

them) and the interests, tastes, and behaviours of their customers. This results in the surveillance of workers and consumers. Because markets are competitive, companies are also interested in monitoring competitors, which has given rise to the phenomenon of industrial espionage. The ideals of modernity (such as the freedom of ownership) also produce phenomena such as income and wealth inequality, poverty, unemployment, precarious living, and working conditions. The establishment of trust, socio-economic differences, and corporate interests are three qualities of modernity that necessitate surveillance. Therefore, modernity, on the one hand, advances the ideal of a right to privacy, but, on the other hand, it must continuously advance surveillance that threatens to undermine privacy rights. An antagonism between privacy ideals and surveillance is therefore constitutive for capitalism.

Workplace surveillance harms employees because the slightest misbehaviour and resistance can be recorded and used for trying to lay them off. Consumer surveillance harms consumers because it enables companies to calculate assumptions about consumers that are error prone, can be used for discriminating between different consumers (based on e.g. income or race) (Gandy 2011), and exploits transaction data and consumer behaviour data that is created by activities of consumers (such as shopping, credit card use, Internet use, etc.) for economic purposes (Fuchs 2011). Economic surveillance is deeply embedded into the antagonisms of capitalism. But also state surveillance is deeply characteristic for modern society. On the one hand, its prevalence can harm citizens by creating a culture of suspicion and fear, in which everybody is seen as an actual or potential criminal or terrorist and the likelihood to be mistaken for engaging in illegal activities is high, on the other hand, state surveillance of companies and the rich could also be used for making power more transparent.

Liberal privacy discourse is highly individualistic, it is always focused on the individual and his/her freedoms. It separates public and private spheres. Privacy in capitalism can best be characterised as an antagonistic value that is one the one side upheld as a universal value for protecting private property, but is at the same time permanently undermined by corporate surveillance into the lives of workers and consumers for profit purposes. Capitalism protects privacy for the rich and companies, but at the same time legitimates privacy violations of consumers and citizens. It thereby undermines its own positing of privacy as universal value.

Given a critical analysis of the privacy concept, the question arises if the concept should best be abolished or if there is another way of coping with its limits. This question will be outlined in the concluding section.

6.5 Conclusion: An Alternative Privacy Concept?

The discussion has shown that the major points of criticism of the modern privacy concept advanced by Marxian analysis are that privacy is frequently fetishised in liberal thought, thereby takes on an ideological character, and tries to mask the negative consequences of capitalism. Marx and Engels have advanced four elements of the critique of the liberal privacy concept that were partly taken up by Arendt and Habermas: (1) privacy as atomism that advances, (2) possessive individualism that harms the public good, (3) legitimises and reproduces the capitalist class structure, and (4) capitalist patriarchy.

Privacy in capitalism protects the rich, companies, and the wealthy. The anonymity of wealth, high incomes, and profits makes income and wealth gaps between the rich and the poor secrets and thereby ideologically helps legitimatising and upholding these gaps. It can therefore be considered an ideological mechanism that helps reproducing and deepening inequality. It would nonetheless be a mistake to fully cancel off privacy rights and to dismiss them as bourgeois values.

I argue for going beyond a bourgeois notion of privacy and to advance a socialist notion of privacy that tries to strengthen the protection of consumers and citizens from corporate surveillance. Economic privacy is therefore posited as undesirable in those cases, where it protects the rich and capital from public accountability, but as desirable, where it tries to protect citizens from corporate surveillance. Public surveillance of the income of the rich and of companies and public mechanisms that make their wealth transparent are desirable for making wealth and income gaps in capitalism visible, whereas privacy protection for workers and consumers from corporate surveillance is also important. In a socialist privacy concept, existing liberal privacy values have therefore to be reversed. Whereas today we mainly find surveillance of the poor and of citizens who are not capital owners, a socialist privacy concept focuses on surveillance of capital and the rich in order to increase transparency and privacy protection of consumers and workers. A socialist privacy concept conceives privacy as a collective right of dominated and exploited groups that need to be protected from corporate domination that aims at gathering information about workers and consumers for accumulating capital, disciplining workers and consumers, and for increasing the productivity of capitalist production and advertising. The liberal conception and reality of privacy as an individual right within capitalism protects the rich and the accumulation of ever more wealth from public knowledge. A socialist privacy concept as a collective right of workers and consumers can protect humans from the misuse of their data by companies. The question therefore is: privacy for whom?

Privacy for dominant groups in regard to secrecy of wealth and power can be problematic, whereas privacy at the bottom of the power pyramid for consumers and normal citizens can be a protection from dominant interests. Privacy rights should therefore be differentiated according to the position people and groups occupy in the power structure. The socialist privacy concept is a form of the RALC because it provides different zones of privacy for different kinds of actors. The differentiation of privacy rights is based on the assumption that the powerless need to be protected from the powerful. Example measures for socialist privacy protection in the area of Internet policies are legal requirements that online advertising must always be based on opt-in options, the implementation and public support of corporate watchdog platforms, the advancement and public support of alternative non-commercial Internet platforms (Fuchs 2012).

Helen Nissenbaum (2010) defines privacy as contextual integrity, which is a heuristic that analyses changes of information processes in specific contexts and flags departures from entrenched privacy practices as violations of contextual integrity. It then analyses if these new practices have moral superiority and if the privacy violation is therefore morally legitimate (Nissenbaum 2010, 164, 182f). In relation to the economy, the concept of contextual integrity helps understanding that privacy plays another role in a context like friendship than in an employment relationship: sharing information about very personal details about your life (like intimacy, sexuality, health, etc.) with a partner or close friends must be judged with other norms than the sharing of the same information with a boss because the first relation is based on close affinity, trust, and feelings of belonging together, whereas the second is based on an economic power relationship. Differentiated values are therefore needed for assessing privacy in both contexts. The concept of socialist privacy is a specific contextualisation of privacy within the economic context – it is a contextualised privacy context, a double contextualisation of privacy: On the one hand, it takes into account the power relationships of the economy and on the other hand, it must in the context of the modern economy take into account class relationships, i.e. the asymmetric power structure of the capitalist economy, in which employers and companies have the power to determine and control many aspects of the lives of workers and consumers. Given the power of companies in the capitalist economy, economic privacy needs to be contextualised in a way that protects consumers and workers from capitalist control and at the same time makes corporate interests and corporate power transparent.

Etzioni (1999) stresses that liberal privacy concepts typically focus on privacy invasions by the state, but ignore privacy invasions by companies. The contemporary undermining of public goods by overstressing privacy rights would not be caused by the state, but rather stem "from the quest for profit by some private companies. Indeed, I find that these corporations now regularly amass detailed accounts about many aspects of the personal lives of millions of individuals, profiles of the kind that until just a few years ago could be compiled only by the likes of the East German Stasi. [...] Consumers, employees, even patients and children have little protection from marketeers, insurance companies, bankers, and corporate surveillance".

(Etzioni 1999, 9f)

The task of a socialist privacy conception is to go beyond the focus of privacy concepts as protection from state interference into private spheres, but to identify those cases, where political regulation is needed for the protection of the rights of consumers and workers.

It is time to break with the liberal tradition in privacy studies and to think about alternatives. The Swedish socialist philosopher Torbjörn Tännsjö (2010) stresses that liberal privacy concepts imply "that one can not only own self and personal things, but also means of production" and that the consequence is "a very closed society, clogged because of the idea of business secret, bank privacy, etc" (Tännsjö 2010, 186). Tännsjö argues that power structures should be made transparent and not be able to hide themselves and operate secretly protected by privacy rights. He imagines based on utopian socialist ideas an open society that is democratic and fosters equality so that (Tännsjö 2010, 191–198) in a democratic socialist society, there is, as Tännsjö indicates, no need for keeping power structures secret and therefore no need for a liberal concept of privacy. However, this does in my view not mean that in a society that is shaped by participatory democracy, all forms of privacy vanish. There are some human acts and situations, such as defecation (Moore 1984), in which humans tend to want to be alone. Many humans would both in a capitalist and a socialist society feel embarrassed having to defecate next to others, for example by using toilets that are arranged next to each other without separating walls. So solitude is not a pure ideology, but to a certain desire also a human need that should be guaranteed as long as it does not result in power structures that harm others. This means that it is necessary to question the liberal-capitalist privacy ideology, to struggle today for socialist privacy that protects workers and consumers, limits the right and possibility of keeping power structures secret, and makes these structures transparent. In a qualitatively different society, we require a qualitatively different concept of privacy, but not the end of privacy. Torbjörn Tannsjö's work is a powerful reminder that it is necessary not to idealise privacy, but to think about its contradictions and its relation to private property.

Conclusion: An Alternative Privacy Concept?

Note

1 Solove (2008) argues that the family, the body, sex, the home, and communications are contexts of privacy.

References

Allen, Anita. 1988. *Uneasy Access*. Totowa, NJ: Rowman & Littlefield.

Arendt, Hannah. 1958. *The Human Condition*. Chicago, IL: University of Chicago Press. Second edition.

Benn, Stanley I. 1971/1984. Privacy, Freedom, and Respect for Persons. In *Philosophical Dimensions of Privacy*, ed. Ferdinand David Schoeman, 223–244. Cambridge, MA: Cambridge University Press.

Bennett, Colin. 2008. *The Privacy Advocates*. Cambridge, MA: MIT Press.

Bennett, Colin and Charles Raab. 2006. *The Governance of Privacy*. Cambridge, MA: MIT Press.

Bloustein, Edward J. 1964/1984. Privacy as an Aspect of Human Dignity. In *Philosophical Dimensions of Privacy*, ed. Ferdinand David Schoeman, 156–202. Cambridge, MA: Cambridge University Press.

Bok, Sissela 1983. *Secrets: On the Ethics of Concealment and Revelation*. New York: Pantheon.

Burk, Dan L. 2007. Privacy and Property in the Global Datasphere. In *Technology Ethics: Cultural Perspectives*, ed. Soraj Hongladarom and Charles Ess, 94–107. Hershey: Idea Group Reference.

Capurro, Rafael. 2005. Privacy. An Intercultural Perspective. *Ethics and Information Technology* 7 (1): 37–47.

Dussel, Enrique. 2008. The Discovery of the Category of Surplus Value. In *Karl Marx's Grundrisse: Foundations of the Critique of the Political Economy 150 Years Later*, ed. Marcello Musto, 67–78. New York: Routledge.

Engels, Friedrich. 1891. The Origin of the Family, Private Poperty and the State. In *Selected Works in One Volume: Karl Marx and Friedrich Engels*, 430–558. London: Lawrence & Wishart.

Ess, Charles. 2005. "Lost in Translation"?: Intercultural Dialogues on Privacy and Information Ethics. *Ethics and Information Technology* 7 (1): 1–6.

Etzioni, Amitai. 1999. *The Limits of Privacy*. New York: Basic Books.

Fried, Charles. 1968/1984. Privacy. In *Philosophical Dimensions of Privacy*, ed. Ferdinand David Schoeman, 203–222. Cambridge, MA: Cambridge University Press.

Froomkin, A. Michael. 2000. The Death of Privacy? *Stanford Law Review* 52: 1461–1543.

Fuchs, Christian. 2012. The Political Economy of Privacy on Facebook. *Television & New Media* 13 (2): 139–159.

Fuchs, Christian. 2011. Critique of the Political Economy of Web 2.0 Surveillance. In *Internet and Surveillance. The Challenges of Web 2.0 and Social Media*, ed. Christian Fuchs, Kees Boersma, Anders Albrechtslund and Kees Boersma, 31–70. New York: Routledge.

Fuchs, Christian. 2003a. Some Implications of Pierre Bourdieu's Works for a Theory of Social Self-Organization. *European Journal of Social Theory* 6 (4): 387–408.

Fuchs, Christian. 2003b. Structuration Theory and Self-Organization. *Systemic Practice and Action Research* 16 (4): 133–167.

Gandy, Oscar. 2011. *Coming to Terms with Chance*. Farnham: Ashgate.

Gavison, Ruth. 1980. Privacy and the Limits of Law. *Yale Law Journal* 89 (3): 421–471.

Gerstein, Robert S. 1978/1984. Intimacy and Privacy. In *Philosophical Dimensions of Privacy*, ed. Ferdinand David Schoeman, 265–271. Cambridge, MA: Cambridge University Press.

Giddens, Anthony. 1984. *The Constitution of Society*. Berkeley: University of California Press.

Gilliom, John. 2001. *Overseers of the Poor*. Chicago, IL: University of Chicago Press.

Gormley, Ken. 1992. One Hundred Years of Privacy. *Wisconsin Law Review* 1992: 1335–1441.

Habermas, Jürgen. 1989. *The Structural Transformation of the Public Sphere*. Cambridge, MA: MIT Press.

Habermas, Jürgen. 1987. *The Theory of Communicative Action. Volume 2*. Boston, MA: Beacon Press.

Hongladarom, Soraj. 2007. Analysis and Justification of Privacy from a Buddhist Perspective. In *Technology Ethics: Cultural Perspectives*, ed. Soraj Hongladarom and Charles Ess, 108–122. Hershey: Idea Group Reference.

Introna, Lucas D. 2000. Privacy and the Computer. Why We Need Privacy in the Information Society. In *Cyberethics*, ed. Robert M. Baird, Reagan Ramsower, and Stuart E. Rosenbaum, 188–199. Amherst, NY: Prometheus.

Lyon, David. 2007. *Surveillance Studies: An Overview*. Cambridge, UK: Polity.

Lyon, David. 2001. *Surveillance Society. Monitoring Everyday Life*. Buckingham: Open University Press.

Lyon, David. 1994. *The Electronic Eye. The Rise of Surveillance Society*. Cambridge: Polity.

Macpherson, Crawford B. 1962. *The Political Theory of Possessive Individualism*. Oxford: Oxford University Press.

Margulis, Stephen. 2003a. On the Status and Contribution of Westin's and Altman's Theories of Privacy. *Journal of Social Issues* 59 (2): 411–430.

Margulis, Stephen. 2003b. Privacy as a Social Issue and Behavioural Concept. *Journal of Social Issues* 59 (2): 243–262.

Marx, Karl. 1867. *Capital. Volume I*. London: Penguin.

Marx, Karl. 1857/58. *Grundrisse*. London: Penguin.

Marx, Karl. 1844. Economic and Philosophic Manuscripts of 1844. In *Economic and Philosophic Manuscripts of 1844 and the Communist Manifesto*, 13–168. Amherst, NY: Prometheus.

Marx, Karl. 1843a. Critique of Hegel's Doctrine of the State. In *Early Writings*, 57–198. London: Penguin.

Marx, Karl. 1843b. On the Jewish Question. In W*ritings of the Young Marx on Philosophy and Society*, 216–248. Indianapolis, IN: Hackett.

Marx, Karl and Friedrich Engels. 1846. *The German Ideology*. Amherst, NY: Prometheus.

Mill, John Stuart. 1965. *Principles of Political Economy. 2 Volumes.* London: University of Toronto Press.

Miller, Arthur. 1971. *The Assault on Privacy.* Cambridge, MA: Harvard University Press.

Moor, James H. 2000. Toward a Theory of Privacy in the InformationAage. In *Cyberethics*, ed. Robert M. Baird, Reagan Ramsower, and Stuart E. Rosenbaum, 200–212. Amherst, NY: Prometheus.

Moore, Barrington. 1984. *Privacy. Studies in Social and Cultural History.* Armonk, NY: M.E. Sharpe.

Negri, Antonio. 1991. *Marx beyond Marx.* London: Pluto.

Nissenbaum, Helen. 2010. *Privacy in Context.* Standford, CA: Stanford University Press.

Nock, Steven. 1993. *The Costs of Privacy: Surveillance and Reputation in America.* New York: de Gruyter.

Posner, Richard A. 1978/1984. An Economic Theory of Privacy. In *Philosophical Dimensions of Privacy*, ed. Ferdinand David Schoeman, 333–345. Cambridge, MA: Cambridge University Press.

Quinn, Michael. 2006. *Ethics for the Information Age.* Boston, MA: Pearson.

Rachel, James. 1975/1984. Why Privacy is Important. In *Philosophical Dimensions of Privacy*, ed. Ferdinand David Schoeman, 290–299. Cambridge, MA: Cambridge University Press.

Rule, James B. 2007. *Privacy in Peril.* Oxford: Oxford University Press.

Schoeman, Ferdinand David. 1992. *Privacy and Social Freedom.* Cambridge: Cambridge University Press.

Schoeman, Ferdinand David. 1984a. Privacy and intimate information. In *Philosophical Dimensions of Privacy*, ed. Ferdinand David Schoeman, 403–418. Cambridge, MA: Cambridge University Press.

Schoeman, Ferdinand David. 1984b. Privacy: Philosophical Dimensions of the Literature. In *Philosophical Dimensions of Pprivacy*, ed. Ferdinand David Schoeman, 1–33. Cambridge, MA: Cambridge University Press.

Schultz, Robert A. 2006. *Contemporary Issues in Ethics and Information Technology.* Hershey, PA: IRM Press.

Sewell, Graham and James Barker. 2007. Neither Good, nor Bad, but Dangerous: Surveillance as an Ethical Paradox. In *The Surveillance Studies Reader*, ed. Sean P. Hier and Josh Greenberg, 354–367. Maidenhead: Open University Press.

Shade, Leslie Regan. 2003. Privacy. In *Encyclopedia of New Media*, ed. Steve Jones, 378–380. Thousand Oaks, CA: Sage.

Shils, Edward. 1966. Privacy: Its Constitution and Vicissitudes. *Law & Contemporary Problems* 31(2): 281–306.

Solove, Daniel J. 2008. *Understanding Privacy.* Cambridge, MA: Harvard University Press.

Solove, Daniel. 2004. *The Digital Person. Technology and Privacy in the Information Age.* New York: New York University Press.

Spinello, Richard. 2006. *CyberEthics: Morality and Law in Cyberspace.* Sudbury, MA: Jones and Bartlett.

Stahl, Bernd Carsten. 2007. Privacy and Security as Ideology. *IEEE Technology and Society Magazine* Spring 2007: 35–45.

Stalder, Felix. 2002. Privacy is not the Antidote to Surveillance. *Surveillance & Society* 1 (1): 120–124.

Tännsjö, Torbjörn. 2010. *Privatliv*. Lidingö: Fri Tanke.

Tavani, Herman T. 2008. Informational Privacy: Concepts, Theories, and Controversies. In *The Handbook of Information and Computer Ethics*, ed. Kenneth Einar Himma and Herman T. Tavani, 131–164. Hoboken, NJ: Wiley.

Wacks, Raymond. 2010. *Privacy. A Very Short Introduction*. Oxford: Oxford University Press.

Warren, Samuel and Louis Brandeis. 1890. The Right to Privacy. *Harvard Law Review* 4 (5): 193–220.

Wasserstrom, Richard A. 1978/1984. Privacy. Some Arguments and Assumptions. In *Philosophical Dimensions of Privacy*, ed. Ferdinand David Schoeman, 317–332. Cambridge, MA: Cambridge University Press.

Westin, Alan. 1967. *Privacy and Freedom*. New York: Altheneum.

Zureik, Elia and L. Lynda Harling Stalker. 2010. The Cross-Cultural Study of Privacy. In *Surveillance, Privacy and the Globalization of Personal Information*, ed. Elia Zureik, Lynda Harling Stalker, Emily Smith, David Lyon, and Yolane E. Chan, 8–30. Montreal: McGill-Queen's University Press.

Chapter Seven
The Ethics and Political Economy of Privacy on Facebook

7.1 Introduction

Facebook has become one of the world's most accessed Internet platforms. In 2011, it held rank number two in the list of the most accessed websites in the world (data source: alexa.com, accessed on 28 May 2011): 41.7% of the world's Internet users accessed Facebook in the three-month period from 28th February to 28th May 2011. Ten years later, Facebook was the seventh most accessed Internet platform (data source: alexa.com, accessed on 26 March 2021). Given the fact that Facebook is a tremendously successful project, it is an important research task to critically analyse the economic structures and power relations of the platform.

In this chapter, I provide an analysis of the political economy of privacy and surveillance on Facebook, which means that the task is to show how privacy on Facebook is connected to surplus value, exploitation, and class (Dussel 2008, 77; Negri 1991, 74). The privacy notion is a foundation for the discussion shortly discussed in Section 7.2. The political economy of privacy on Facebook is analysed in Section 7.3. Potential alternatives and elements of a socialist privacy strategy are identified in Section 7.4. Finally, some conclusions are drawn in Section 7.5.

7.2 Beyond the Liberal Concept of Privacy

Definitions of informational privacy have in common that they deal with the moral questions of how information about humans should be processed, who shall have access to this data, and how this access shall be regulated (Tavani 2008). They also share the conviction that some form of data protection is needed. Etzioni (1999) stresses that it is

DOI: 10.4324/9781003279488-9

a typical American liberal belief that strengthening privacy can cause no harm and says that privacy can undermine common goods (public safety, public health).

Countries like Switzerland, Liechtenstein, Monaco, and Austria have a tradition of the relative anonymity of bank accounts and transactions. Money as private property is seen as an aspect of privacy, which means that financial information tends to be kept secret and withheld from the public. In Switzerland, the bank secret is defined in the Federal Banking Act (§47). The Swiss Bankers Association sees bank anonymity as a form of "financial privacy"[1] that needs to be protected and speaks of "privacy in relation to financial income and assets".[2] In many countries, information about income and the profits of companies (except for public companies) is treated as a secret, a form of financial privacy. The problem of secret bank accounts/transactions and the non-transparency of richness and company profits is not only that financial privacy can support tax evasion, black money affairs, and money laundering, but also that it hides wealth gaps. Financial privacy reflects the classical liberal account of privacy. So for example John Stuart Mill formulated the right of the propertied class to economic privacy as "the owner's privacy against invasion" (Mill 1965, 232). Economic privacy in capitalism (the right to keep information about income, profits and bank transactions secret) protects the rich, companies, and the wealthy. The anonymity of wealth, high incomes, and profits makes income and wealth gaps between the rich and the poor invisible and thereby ideologically helps legitimatising and upholding these gaps. Financial privacy is an ideological mechanism that helps reproducing and deepening inequality. Karl Marx, who positioned privacy in relation to private property, first formulated the critique of the liberal concept of privacy. The liberal concept of the private individual and privacy would see man as "an isolated monad, withdrawn into himself. [...] The practical application of the right of liberty is the right of private property" (Marx 1843b, 235). Modern society's constitution would be the "constitution of private property" (Marx 1843a, 166). Torbjörn Tännsjö (2010) stresses that liberal privacy concepts imply "that one can not only own one-self and personal things, but also means of production" and that the consequence is "a very closed society, clogged because of the idea of business secret, bank privacy, etc." (Tännsjö 2010, 186; translation from Swedish by the author).

It would nonetheless be a mistake if we were to fully cancel off privacy rights and dismissed them as bourgeois values. Liberal privacy discourse is highly individualistic, it is always focused on the individual and his/her freedoms. It separates the public and the private sphere. Privacy in capitalism can best be characterised as an antagonistic value that is on the one side upheld as a universal value for protecting private property, but is at the same time permanently undermined by corporate and state surveillance into the lives of humans for the purpose of capital accumulation. Capitalism protects privacy for the rich and companies, but at the same time legitimates privacy violations of consumers and citizens. Liberal privacy values have their limit and find their immanent critique within the reality of liberal-capitalist economies.

When discussing privacy on Facebook, we should therefore go beyond a bourgeois notion of Facebook and try to advance a socialist notion of privacy that aims at strengthening the protection of consumers and citizens from corporate surveillance and other forms of domination. Economic privacy should therefore be posited as undesirable in those cases, where it protects the rich and capital from public accountability, but as desirable, where it tries to protect citizens, workers, and consumers from corporate surveillance. Public surveillance of the income of the rich and of companies and public mechanisms that make their wealth transparent are desirable for making wealth and income gaps in capitalism visible, whereas privacy protection from corporate surveillance is at the same time also important. In a socialist privacy concept, the existing privacy values have to be reversed.

Whereas today we mainly find surveillance of the poor, workers, consumers, everyday citizens, and privacy protests private property, in contrast to this reality a socialist privacy concept focuses on surveillance of capital and the rich in order to increase transparency and privacy protection of consumers and workers. A socialist privacy concept conceives privacy as a collective right of dominated and exploited groups that need to be protected from corporate domination that aims at gathering data for accumulating capital, disciplining workers and consumers, and for increasing the productivity of capitalist production and advertising. The liberal conception and reality of privacy as an individual right within capitalism protects the rich and the accumulation of ever more wealth from public knowledge. A socialist privacy concept as a collective right of workers and consumers can protect humans from the misuse of their data by companies. The questions therefore are: For whom should privacy be guaranteed, for whom not? What type of privacy should we struggle for on Facebook? Privacy for dominant groups in regard to the secrecy of wealth and power is problematic, whereas privacy at the bottom of the power pyramid for consumers, workers, and normal citizens is a protection from dominant interests. Privacy rights should therefore be differentiated according to the position people and groups occupy in the power structure. In relation to Facebook, this means that the main privacy topic is not how many information users make available to the public, but rather which user data is used by Facebook

Beyond the Liberal Concept of Privacy

for advertising purposes, in which sense users are exploited in this process and how users can be protected from the negative consequences of economic surveillance on Facebook.

Helen Nissenbaum (2010) argues that one should go beyond the control theory and the access theory of privacy and consider privacy as contextual integrity. Contextual integrity is a heuristic that analyses changes of information processes in specific contexts and flags departures from entrenched privacy practices as violations of contextual integrity. It then analyses if these new practices have moral superiority and if the privacy violation is therefore morally legitimate (Nissenbaum 2010, 164, 182f). Nissenbaum mentions as relevant contexts education, health care, psychoanalysis, voting, employment, the legal system, religion, family, and the marketplace (Nissenbaum 2010, 130, 169–179). Contextual privacy is "preserved when informational norms are respected and violated when informational norms are breached. [...] whether or not control is appropriate depends on the context, the types of information, the subject, sender, and recipient" (Nissenbaum 2010, 140, 148). In relation to the economy, the concept of contextual integrity helps understanding that privacy plays another role in a context like friendship than in an employment relationship: sharing information about very personal details about your life (like intimacy, sexuality, health, etc.) with a partner or close friends must be judged with other norms than the sharing of the same information with a boss because the first relation is based on close affinity, trust, and feelings of belonging together, whereas the second is based on an economic power relationship. Differentiated values are therefore needed for assessing privacy in both contexts. The concept of socialist privacy is a specific contextualisation of privacy within the economic context – it is a contextualised privacy context, a double contextualisation of privacy: On the one hand, it takes into account the power relationships of the economy and on the other hand, it must in the context of the modern economy take into account class relationships, i.e. the asymmetric power structure of the capitalist economy, in which employers and companies have the power to determine and control many aspects of the lives of workers and consumers. Given the power of companies in the capitalist economy, economic privacy needs to be contextualised in a way that protects consumers and workers from capitalist control and at the same time makes corporate interests and corporate power transparent. For privacy on Facebook this means it should be made transparent what data Facebook stores about its users and that users should be protected from Facebook's exploitation of their data for economic purposes. This requires a differentiated concept of economic privacy that distinguishes between the roles of consumers, workers, and companies in the capitalist economy.

Mainstream research about Facebook and social networking sites in general engages in privacy fetishism by focusing on the topic of information disclosures by users (for a critique of such studies see: Fuchs 2009b, Chapter 3). These studies consider privacy threatened because users would disclose too much information about themselves. They stress associated risks. They conceive privacy strictly as an individual phenomenon that can be protected if users behave in the correct way and do not disclose too much information. The moralistic tone in these studies ignores how Facebook commodifies data and exploits users as well as the societal needs and desires underpinning information sharing on Facebook. As a result, this discourse is individualistic and ideological. It focuses on the analysis of individual behaviour without seeing and analysing how this use is conditioned by the societal contexts of information technologies, such as surveillance, the global war against terror, corporate interests, neoliberalism, and capitalist development.

These contexts make it incumbent for critical Internet studies to analyse Facebook privacy in the context of the political economy of capitalism.

7.3 The Political Economy of Facebook

Alvin Toffler (1980) introduced the notion of the prosumer in the early 1980s. It means the "progressive blurring of the line that separates producer from consumer" (Toffler 1980, 267). Toffler describes the age of prosumption as the arrival of a new form of economic and political democracy, self-determined work, labour autonomy, local production, and autonomous self-production. But he overlooks that prosumption is used for outsourcing work to users and consumers, who work without payment. Thereby corporations reduce their investment costs and labour costs, jobs are destroyed, and consumers who work for free are exploited. They produce surplus value that is appropriated and turned into profit by corporations without paying wages. Notwithstanding Toffler's uncritical optimism, his notion of the "prosumer" describes important changes of media structures and practices and can therefore also be adopted for critical studies.

It becomes ever more frequent that users or observers of Facebook argue that Facebook exploits them by making a profit with the help of their data. The concept of exploitation is frequently not explained and clarified in such circumstances. Karl Marx (1867) provided the best and most important explanation of exploitation in capitalism. In order to understand how exploitation on Facebook exactly works and to avoid that the use of the exploitation concept not just is a moral appeal, but a critique that is analytically grounded, it is therefore necessary to go into some details of the Marxian political economy.

The Political Economy of Facebook

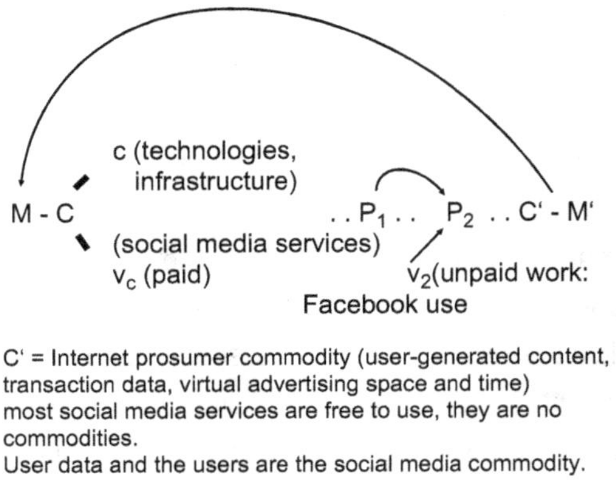

C' = Internet prosumer commodity (user-generated content, transaction data, virtual advertising space and time) most social media services are free to use, they are no commodities.
User data and the users are the social media commodity.

FIGURE 7.1 Capital accumulation on Facebook.

Figure 7.1 shows the process of capital accumulation on Facebook. Facebook invests money (M) for buying capital: technologies (server space, computers, organisational infrastructure, etc.) and labour power (paid Facebook employees). These are the constant capital (c) and the variable capital (v1) outlays. The outcome of the production process P1 is not a commodity that is directly sold, but rather social media services (the Facebook platform) that are made available without payment to users. The Facebook employees, who create the Facebook online environment that is accessed by Facebook users, produce part of the surplus value. The Facebook users make use of the platform for generating content that they upload (user-generated data). The constant and variable capital invested by Facebook (c, v1) that is objectified in the Facebook environment is the prerequisite for their activities in the production process P2. Their products are user-generated data, personal data, and transaction data about their browsing behaviour and communication behaviour on Facebook. They invest a certain labour time v2 in this process. Facebook sells the users' data commodity to advertising clients at a price that is larger than the invested constant and variable capital. The surplus value contained in this commodity is partly created by the users, partly by the Facebook employees. The difference is that the users are unpaid and therefore infinitely exploited. Once the Internet prosumer commodity that contains the user-generated content, transaction data, and the right to access virtual advertising space and time is sold to advertising clients, the commodity is transformed into money capital and surplus value is realised into money capital.

For Marx (1867), the profit rate is the relation of profit to investment costs:

$p = s/(c + v)$ = surplus value/(constant capital [=fixed costs] + variable capital [=wages]).

If Internet users become productive web 2.0 prosumers, then in terms of Marxian class theory this means that they become productive labourers, who produce surplus value and are exploited by capital because for Marx productive labour generates a surplus. Therefore the exploitation of surplus value in the case of Facebook is not merely accomplished by those who are employed for programming, updating, and maintaining the soft- and hardware, performing marketing activities, and so on, but by them, the users, and the prosumers that engage in the production of user-generated content. New media corporations do not (or hardly) pay the users for the production of content. A widely used accumulation strategy is to give the users free access to services and platforms, let them produce content, and to accumulate a large number of prosumers that are sold as a commodity to third-party advertisers. Not a product is sold to the users, but the users are sold as a commodity to advertisers. The more users a platform has, the higher the advertising rates can be set. The productive labour time that is exploited by capital, on the one hand, involves the labour time of the paid employees and, on the other hand, all of the time that is spent online by the users. For the first type of knowledge labour, new media corporations pay salaries. The second type of knowledge is produced completely for free. There are neither variable nor constant investment costs. The formula for the profit rate needs to be transformed for this accumulation strategy:

$p = s / (c + v1 + v2)$, s ... surplus value, c ... constant capital, v1 ... wages paid to fixed employees, v2 ... wages paid to users.

The typical situation is that $v2 \geq 0$ and that v2 substitutes v1. If the production of content and the time spent online were carried out by paid employees, the variable costs would rise and profits would therefore decrease. This shows that prosumer activity in a capitalist society can be interpreted as the outsourcing of productive labour to users, who work completely for free and help maximising the rate of exploitation ($e = s/v$, = surplus value/variable capital) so that profits can be raised and new media capital may be accumulated. This situation is one of infinite exploitation of the users. Capitalist prosumption is an extreme form of exploitation, in which the prosumers work completely for free.

What does it mean that Facebook prosumers work for free and are exploited? Adam is a 13-year-old school kid and heavy Facebook user. He has 2,000 Facebook friends, writes 50 wall postings a day, interacts with at least 40 of his close contacts and colleagues over Facebook a day, updates his status at least ten times a day, uploads a lot of commented videos and photos from each of his weekends that he tends to spend together with his girlfriend in the countryside. Yet there is one thing that puzzles him: The advertisements at the right-hand side of his profile frequently have to do with what he has done the last weekend or he intends to do the next weekend. Adam wonders how this comes and does not feel so well about the fact that obviously his personal data is used for economic ends and that he does not exactly know and cannot control which of his data and usage behaviour is stored, assessed, and sold. The answer is that Facebook closely monitors all of his contacts, communications, and data and sells this information to companies, who provide targeted advertisements to Adam. Facebook thereby makes a lot of profit and could not exist without the unpaid labour that Adam and millions of other of his fellow Facebook workers conduct. Adam is the prototypical Facebook child worker.

Dallas Smythe (1981/2006) suggests that in the case of media advertisement models, the audience is sold as a commodity to advertisers: "Because audience power is produced, sold, purchased and consumed, it commands a price and is a commodity. [...] You audience members contribute your unpaid work time and in exchange you receive the program material and the explicit advertisements" (Smythe 1981/2006, 233, 238; see also Smythe 1977).

Smythe's audience commodity hypothesis has resulted in a sustained debate (see for example, Bolin 2005, 2009; Hearn 2010; Hesmondhalgh 2010; Jhally 1987; Jhally and Livant 1986; Lee 2011; Livant 1979; Manzerolle 2010; Meehan 1993; Murdock 1978; Smythe 1978), including a critique by Jhally and Livant (1986) that not the audience, but their watching time is the commodity. Both Jhally/Livant's approach (Andrejevic 2002) and Smythe's approach (Fuchs 2009a, 2010a, 2010c, 2011b) remain important today for discussing commodification on the Internet and share a focus on commodification and exploitation.

With the rise of user-generated content, free-access social networking platforms, and other ad-based platforms, the web seems to come closer to TV or radio in their accumulation strategies. The users who upload photos, and images, write wall posting and comments, send mail to their contacts, accumulate friends or browse other profiles on Facebook, constitute an audience commodity that is sold to advertisers. The difference between the audience commodity on traditional mass media and on the Internet is that in the latter case the users are also content producers; there is user-generated content, the users engage in permanent creative activity, communication, community building, and content production. In the case of Facebook, the audience commodity is an Internet prosumer commodity.

Surveillance of Facebook prosumers occurs via corporate web platform operators and third-party advertising clients, which continuously monitor and record personal data and online activities. Facebook surveillance creates detailed user profiles so that advertising clients know and can target the personal interests and online behaviours of the users. Facebook sells its prosumers as a commodity to advertising clients; their exchange value is based on permanently produced use values, i.e. personal data and interactions.

Facebook prosumers are double objects of commodification: they are first commodified by corporate platform operators, who sell them to advertising clients, and this results, second, in an intensified exposure to commodity logic. They are permanently exposed to commodity propaganda presented by advertisements while they are online. Most online time is advertising time.

The labour side of the capital accumulation strategy of social media corporations is digital playbour. Kücklich (2005) first introduced the term playbour (play+labour) and in the meantime conferences like "Digital Labour: Workers, Authors, Citizens" (University of Western Ontario 2009) and "The Internet as Playground and Factory" (New School 2009) have helped to advance the discourse about digital playbour. The exploitation of digital playbour is based on the collapse of the distinction between work time and playtime. In the Fordist mode of capitalist production, work time was the time of pain and the time of repression and surplus repression of the human drive for pleasure; whereas leisure time was the time of Eros (Marcuse 1955). In contemporary capitalism, play and labour, Eros and Thanatos, the pleasure principle and the death drive partially converge: workers are expected to have fun during work time and play time becomes productive and work-like. Playtime and work time intersect and all human time of existence tends to be exploited for the sake of capital accumulation. The exploitation of Facebook labour is one expression of these changes of capitalist production and the corresponding transformation of the structure of drives.

Arendt (1958) and Habermas (1989) stress that capitalism has traditionally been based on a separation of the private and the public sphere. Facebook is a typical manifestation of a stage of capitalism, in which the relation of the public and the private and labour and play collapses and in which this collapse is exploited by capital. "The distinction between the private and the public realms [...] equals the distinction between things that should be shown and things that should not be hidden" (Arendt 1958, 72). On Facebook, all private data and user behaviour is shown to the corporation, which commodifies both, whereas it is hidden from the users what exactly happens with their data and to whom these data are sold for the task of targeting advertising. So the main form of privacy on Facebook is the intransparency of capital's use of personal user data that is based on the private appropriation of user data by Facebook. The private user dimension of Facebook

The Political Economy of Facebook

is that content is user-generated by individual users. When it is uploaded to Facebook or other social media, parts of it (to a larger or smaller degree depending on the privacy settings the users choose) become available to lots of people, whereby the data obtains a more public character. The public availability of data can both have advantages (new social relations, friendships, staying in touch with friends, family, relatives over distance, etc.) and disadvantages (job-related discrimination, stalking, etc.) for users (Fuchs 2009b, 2010b, 2010d). The private–public relation has another dimension on Facebook: the privately generated user data and the individual user behaviour become commodified. Both types of data are sold to advertising companies so that targeted advertising is presented to users and Facebook accumulates profit that is privately owned by the company. Facebook commodifies private data that is used for public communication in order to accumulate capital that is privately owned. The users are excluded from the ownership of the resulting money capital, i.e. they are exploited by Facebook and are not paid for their creation of surplus value (Fuchs 2010a). Facebook is a huge advertising-, capital accumulation-, and user exploitation-machine. Data surveillance is the means for Facebook's economic ends. Facebook permanently monitors users for economic ends, which means that no economic privacy is guaranteed to them: It remains unknown to users, which of their information and user behaviour is exactly used for targeted advertising, they cannot control the use of their data and usage behaviour data and are not protected from the commodification of their personal data.

The use of targeted advertising and economic surveillance is legally guaranteed by Facebook's privacy policy. Facebook's privacy policy is a typical expression of a self-regulatory privacy regime, in which businesses define largely themselves how they process personal user data. In general, US data protection laws only cover government databanks, leaving commercial surveillance untouched in order to maximise profitability (Ess 2009, 56; Lyon 1994, 15; Rule 2007, 97; Zureik 2010, 351). Facebook's terms of use and its privacy policy are characteristic for this form of self-regulation. When privacy regulation is voluntary, the number of organisations protecting the privacy of consumers tends to be very small (Bennett and Raab 2006, 171).

7.4 Socialist Privacy Ideals and Social Networking

I question privacy concepts that protect and keep secret capitalist interests and socio-economic inequality. This means that the liberal privacy concept should be challenged by a socialist privacy concept that protects workers and consumers. On Facebook, the "audience" is a worker/consumer – a prosumer. How can socialist privacy protection policies on Facebook look like? One basic insight of this chapter is that privacy protection of consumers, prosumers, and workers can only be achieved in an economy that is not ruled by profit interests, but controlled and managed by prosumers, consumers, and producers, which would at the same time mean the end of the need for forms of privacy that protect us from domination. If there were no profit motive on Internet platforms, then there would be no need to commodify the data and usage behaviour of Internet users. Achieving such a situation is however not primarily a technological task, but one that requires changes of society.

True privacy of consumers, workers, and prosumers is only possible in a participatory democracy. There are today many claims that the Internet has with the emergence of "social media" and "web 2.0" become "participatory". So for example Henry Jenkins argues that with the emergence of a convergence culture "the Web has become a site of consumer participation" (Jenkins 2008, 137) or Axel Bruns (2008, 227f) says that Flickr, YouTube, MySpace, and Facebook are environments of "public participation". Such accounts do not take into account the socialist origins of the concept of participatory democracy.

Staughton Lynd introduced the notion of participatory democracy to the academic debate in 1965. Lynd (1965) used the term for describing the organisation principle of the Students for a Democratic Society (Lynd 1965). Held (1996, 271) says that a key feature of participatory democracy is the "direct participation of citizens in the regulation of the key institutions of society, including the workplace and local community". The two most important thinkers of participatory democracy theory are Crawford Brough Macpherson (1973) and Carole Pateman (1970), who are both socialist thinkers. Participatory democracy means that "democratic rights need to be extended from the state to the economic enterprise and the other central organizations of society" (Held 1996, 268). Some of the central principles of participatory democracy are the following ones (for a full discussion see: Fuchs 2011b, Chapter 7):

1) The intensification and extension of democracy into all realms of life: Not just the political systems, but all realms of life – including the economy – are considered as being systems of power that require democracy in order to be just.

2) Developmental powers as the essence of man and the maximisation of human Man's essence is understood as consisting of a number of positive capacities (development powers, such as co-operation, sociality, emotional activities, etc.) that can depending on the power structures of society be developed to certain extents.

Socialist Privacy Ideals and Social Networking

3) Extractive power as an impediment for participatory democracy: Macpherson (1973) argues that capitalism is based on an exploitation of human powers that limits the development of human capacities because the modern economy

"by its very nature compels a continual net transfer of part of the power of some men to others [for the benefit and the enjoyment of the others], thus diminishing rather than maximizing the equal individual freedom to use and develop one's natural capacities".

(Macpherson 1973, 10–11)

4) Participatory economy: A central aspect of a participatory democracy is a democratic economy, which requires a "change in the terms of access to capital in the direction of more nearly equal access" (Macpherson 1973, 71) and "a change to more nearly equal access to the means of labour" (73). Pateman (1970) terms the grassroots organisation of firms and the economy in a participatory democracy "self-management". A self-managed economy does not consist of classes, there is no need for one class to control and monitor the activities of another class in order to protect and maintain its hegemony. Therefore there is no need for economic privacy violations. Social networking sites and other Internet platforms need to be controlled by the users themselves and organised within the framework of a participatory economy in order to be sensitive for the economic privacy of users.

The overall goal of socialist Internet privacy politics is to drive back the commodification of user data and the exploitation of prosumers by advancing the decommodification of the Internet. Three strategies for achieving this goal are the advancement of opt-in on-line advertising, civil society surveillance of Internet companies, and the establishment and support of alternative platforms.

7.4.1 Opt-in Privacy Policies

Oscar Gandy argues that an alternative to opt-out solutions of targeted advertising

is opt-in solutions that are based on the informed consent of consumers. When individuals

"wish information or an information-based service, they will seek it out. IT is not unreasonable to assume that individuals would be the best judge of when they are the most interested and therefore most receptive to information of a particular kind. Others with information to provide ought to assume that, unless requested, no information is desired. This would be the positive option. Through a variety of means, individuals would provide a positive indication that yes, I want to learn, hear, see more about this subject at this time. Individuals should be free to choose when they are ready to enter the market for information".

(Gandy 1993, 220)

"The value in the positive option is its preservation of the individual's right to choose" (Gandy 1993, 221). Culnan and Bies (2003) argue that opt-in is a form of procedural justice and a fair information practice.

Opt-in privacy policies are typically favoured by consumers and data protectionists, whereas companies and marketing associations prefer opt-out and self-regulation advertising policies in order to maximise profit (Bellman et al. 2004; Federal Trade Commission 2000; Gandy 1993; Quinn 2006; Ryker et al. 2002; Starke-Meyerring and Gurak 2007). Socialist privacy legislation could require all commercial Internet platforms to use advertising only as an opt-in option, which would strengthen the users' possibility for self-determination. Within capitalism, forcing corporations by state laws to implement opt-in mechanisms is certainly desirable, but at the same time it is likely that corporations will not consent to such policies because opt-in is likely to reduce the actual amount of surveilled and commodified user data significantly, which results in a drop of advertising profits. Organising targeted advertising as opt-in instead of as opt-out (or no-option) does not establish economic user privacy, but is a step toward strengthening the economic privacy of users.

7.4.2 Corporate Watch-Platforms as Form of Struggle against Corporate Domination

In order to circumvent the large-scale surveillance of consumers, producers, and consumer–producers, movements and protests against economic surveillance are necessary. Kojin Karatani (2005) argues that consumption is the only space in capitalism, where workers become subjects that can exert pressure by consumption boycotts on capital. I do not think that this is correct because also strikes show the subject position of workers that enables them to boycott production, to cause financial harm to capital, and to exert pressure in order to voice political demands. However, Karatani in my opinion correctly argues that the role of the consumer has been underestimated in Marxist theory and practice. That in the contemporary media landscape media consumers become media producers who work and create surplus value shows the importance of the role of consumers in contemporary capitalism and of "the transcritical moment where workers and consumers intersect" (Karatani 2005, 21). For political strategies, this brings up the actuality of an associationist movement that is "a transnational association of consumers/workers" (Karatani 2005, 295) that engages in "the class struggle against capitalism" of "workers qua consumers or consumers qua workers" (Karatani 2005, 294).

Critical citizens, critical citizens' initiatives, consumer groups, social movement groups, critical scholars, unions, data protection specialists/groups, consumer protection specialists/groups, critical politicians, and critical political parties should observe closely the surveillance operations of corporations and document these mechanisms and instances, where corporations and politicians take measures that threaten privacy or increase the surveillance of citizens. Such documentation is most effective if it is easily accessible to the public. The Internet provides means for documenting such behaviour. It can help to watch the watchers and to raise public awareness. In recent years, corporate watch organisations that run online watch platforms have emerged.
Examples of corporate watch organisations are as follows:

CorpWatch Reporting (http://www.corpwatch.org),

Transnationale Ethical Rating (http://www.transnationale.org),

The Corporate Watch Project (http://www.corporatewatch.org),

Multinational Monitor (http://www.multinationalmonitor.org),

Corporate Crime Reporter (http://www.corporatecrimereporter.com),

Corporate Europe Observatory (http://www.corporateeurope.org),

Corporate Critic Database (http://www.corporatecritic.org).

Transnationale Ethical Rating aims at informing consumers and research about corporations. Its ratings include quantitative and qualitative data about violations of labour rights, violations of human rights, layoff of employees, profits, sales, earnings of CEOs, boards, president and managers, financial offshoring operations, financial delinquency, environmental pollution, corporate corruption, and dubious communication practices. Dubious communication practices include an "arguable partnership, deceptive advertising, disinformation, commercial invasion, spying, mishandling of private data, biopiracy and appropriation of public knowledge" (http://www.transnationale.org/aide.php, accessed on 26 March 2021). The topics of economic privacy and surveillance are here part of a project that wants to document *corporate social irresponsibility*. Privacy is not the only topic, but can on corporate watch platforms be situated in the larger political-economic context of corporate social irresponsibility (the counterpart of the CSR ideology).

On the one hand, it is important to document and gather data about the corporate irresponsibility of Internet corporations. On the other hand, it looks like these data are not very complete and not many Internet corporations are thus far included. So one could for example also document Google's targeted advertising practices. Onlien corporations' data storage and usage practices are highly intransparent to users and leave it unclear for the single user, which data about her/him is exactly stored and commodified. In any case, more efforts are required in order to advance the documentation of corporate social irresponsibility of Internet corporations and to contextualise privacy violations within the process of watching the watchers.

There is a difference between the surveillance of prosumers by Internet corporations and the process of watching the corporate watchdogs. The first process is a process of data collection about users as part of the attempt to exploit users and to thereby deepen the class power of one class at the expense of another one. Corporate watch platforms on the other hand are attempts by those resisting against asymmetric economic power relations to struggle against the powerful class of corporations by documenting data that should make economic power transparent. Prosumers and their data can only be made visible, but not transparent. There is a difference between surveillance for erecting visibility over oppressed groups, which is the attempt to control and further oppress them, and the attempt of making the powerful transparent, which is a self-defence mechanism and form of struggle of the oppressed in order to try to defend themselves against oppression. "'Surveillance' suggests the operation of authority, while 'transparency' suggest the operation of democracy, of the powerful being held accountable" (Johnson and Wayland 2010, 25). Johnson and Wayland (2010) point out that the notion of transparency should be used in relation to economic and political power.

Also, WikiLeaks is a mechanism that tries to make power transparent by leaking secret documents about political and economic power. WikiLeaks does not itself engage in collecting information about the powerful, but relies on anonymous online submissions by insiders, who realise the wrongdoings of institutions and want to contribute to more transparency of what is actually happening. Its overall goal is to make power transparent:

> "WikiLeaks is a multi-national media organization and associated library. It was founded by its publisher Julian Assange in 2006. WikiLeaks specializes in the analysis and publication of large datasets of censored or otherwise restricted

official materials involving war, spying and corruption. It has so far published more than 10 million documents and associated analyses".
(https://wikileaks.org/What-is-WikiLeaks.html, accessed on 26 March 2021)

WikiLeaks and corporate watch platforms have in common that they are both Internet projects that try to make powerful structures transparent as part of the struggle against powerful institutions. The prosecution and imprisonment of Julian Assange have made WikiLeaks' work difficult.

There are no easy solutions to the problem of civil rights limitations due to electronic surveillance. Opting out of existing advertising options is not a solution to the problem of economic and political surveillance. Even if users opt-out, media corporations will continue to collect, assess, and sell personal data, to sell the users as audience commodity to advertising clients, and to give personal data to the police. To try to advance critical awareness and to surveil corporate and political surveillers are important political moves for guaranteeing civil rights, but they will ultimately fail if they do not recognise that electronic surveillance is not a technological issue that can be solved by technological means or by different individual behaviours, but only by bringing about changes of society. Therefore the topic of electronic surveillance should in the public debate be situated in the context of larger societal problems.

7.4.3 Alternative Internet Platforms

The third strategy of socialist privacy politics is to establish and support non-commercial, non-profit Internet platforms. It is not impossible to create successful non-profit Internet platforms, as the example of Wikipedia, which is advertising-free, provides free access, and is financed by donations, shows. The most well-known alternative social networking site project is Diaspora, which tries to develop an open-source alternative to Facebook. It is a project created in late 2010 by the four NYU students Dan Grippi, Maxwell Salzberg, Raphael Sofaer, and Ilya Zhitomirskiy.

Diaspora defines itself as "privacy-aware, personally controlled, do-it-all, open source social" (http://www.joindisaspora.com, accessed on 11 November 2010). It is not funded by advertising, but by donations. Three design principles of Diaspora are choice, ownership, and simplicity:

"Choice: Diaspora lets you sort your connections into groups called aspects. Unique to Diaspora, aspects ensure that your photos, stories and jokes are shared only with the people you intend. Ownership: You own your pictures, and you shouldn't have to give that up just to share them. You maintain ownership of everything you share on Diaspora, giving you full control over how it's distributed. Simplicity: Diaspora makes sharing clean and easy – and this goes for privacy too. Inherently private, Diaspora doesn't make you wade through pages of settings and options just to keep your profile secure".

(http://www.joindisaspora.com, accessed on 21 March 2011)

The Diaspora team is critical of the control of personal data by corporations. It describes Facebook as "spying for free" and the activities of Facebook and other corporate Internet platforms in the following way. Maxwell Salzberg: "When you give up that data, you're giving it up forever. [...] The value they give us is negligible in the scale of what they are doing, and what we are giving up is all of our privacy" (http://www.nytimes.com/2010/05/12/nyregion/12about.html). Ilya Zhitomirskiy:

"For the features that we get on blogs, social networks and social media sites, we sacrifice lots of privacy. [...] The features that we get are not anything special. [...] What will happen [...] when one of these big large companies just goes bust, but has as one of its assets all of your personal data and all of our personal data, our communications, our photos, our comments? [...] They are in power to do what they please with it".

(http://vimeo.com/11242604)

The basic idea of Diaspora is to circumvent the corporate mediation of sharing and communication by using decentralised nodes that store data that is shared with friends (http://vimeo.com/11242736). Each user has his/her own data node that s/he fully controls. Maxwell Salzberg: "Sharing is a human value. Sharing makes the Internet really awesome" (http://vimeo.com/11099292). [...] Ilya Zhitomirskiy: On Diaspora, users are no longer dependent on "corporate networks, who want to tell you that sharing and privacy are mutually exclusive" (http://vimeo.com/11099292).

Diaspora aims to enable users to share data with others and at the same time to protect them from corporate domination and from having to sacrifice their data to corporate purposes in order to communicate and share. Diaspora can therefore be considered as a socialist Internet project that practically tries to realise a socialist privacy concept. The Diaspora team is inspired by the ideas of Eben Moglen, author of the *dotCommunist Manifesto*. He says that an important political goal and possibility today is the "liberation of information from the control of ownership" with the help of networks that are "based on association among peers without hierarchical control, which replaces the coercive system"

Socialist Privacy Ideals and Social Networking

(Moglen 2003) of capitalist ownership of knowledge and data. "In overthrowing the system of private property in ideas, we bring into existence a truly just society, in which the free development of each is the condition for the free development of all" (Moglen 2003).

Ten years after its creation, Diaspora still exits and describes itself as the "online social world where you are in control" and that is based on the three principles of decentralisation, freedom, and privacy (https://diasporafoundation.org/, accessed on 26 March 2021). Diaspora has not been able to compete with Facebook, the world's dominant social networking site. While Facebook has billions of users that it reaches via Facebook, Instagram, and WhatsApp, Diaspora in March 2021 had around 750,000 registered users (data source: https://the-federation.info/diaspora, accessed on 26 March 2021). The problem of alternative projects, media, and platforms in capitalist society is that they often lack people, money, attention, reputation, paid labour-time, influence, resources, and other forms of power. Not-for-profit projects are in capitalism fair, nice, just, and alternative, but they often cannot compete with and properly question and struggle against corporate giants. They are often built on voluntary, self-exploitative, unpaid, or very low-paid labour and resource precarity. Alternative media, platforms, and projects have difficulties challenging capitalist corporations.

There are diffuse feelings of discontent of many users with Facebook's privacy practices that have manifested themselves into groups against the introduction of Facebook Beacon, news feed, mini-feed, etc., the emergence of the web 2.0 suicide machine (http://suicidemachine.org/), or the organisation of a Quit Facebook Day. These activities are mainly based on liberal and Luddite ideologies, but if they were connected to ongoing class struggles against neoliberalism (like the ones' of students throughout the world in the aftermath of the new global capitalist crisis) and the commodification of the commons, they could grow in importance. Existing struggles could be connected to the attempts to establish opt-in policies, corporate social media watchdogs, and alternative social media. Another idea is a campaign that demands that Facebook and all other corporate social media platforms to pay a wage to its users. On the one hand, such a campaign could create attention for the exploitation of user labour, on the other hand, its goal (a wage paid by corporate social media providers) would be short-sighted if it did not aim at the same time at overcoming the wage economy and exploitation as such. The crisis has created the conditions for new struggles, but the main reaction of the people is that in many countries there is a shift toward the right and extreme-right and the rise of hyper-neoliberalism. Besides its strong objective foundations, class struggle from below as part of socialist strategy today is only "latent or manifests itself only in isolated and sporadic phenomena" (Marx 1867, 96).

7.5 Conclusion

Facebook founder and CEO Mark Zuckerberg says that Facebook is about the "concept that the world will be better if you share more" (*Wired*, August 2010). Zuckerberg has repeatedly said that he does not care about profit, but wants to help people with Facebook's tools and wants to create an open society. Kevin Colleran, Facebook advertising sales executive, says in the Wired story that "Mark is not motivated by money". In a Times story,[3] Zuckerberg said:

> "The goal of the company is to help people to share more in order to make the world more open and to help promote understanding between people. The long-term belief is that if we can succeed in this mission then we also [will] be able to build a pretty good business and everyone can be financially rewarded. [...] The Times: Does money motivate you? Zuckerberg: No".

If Zuckerberg really does not care about profit, why is Facebook then not a non-commercial platform and why does it use targeted advertising? The problems of targeted advertising are that it aims at controlling and manipulating human needs, that users are normally not asked if they agree to the use of advertising on the Internet, but have to agree to advertising if they want to use commercial platforms (lack of democracy), that advertising can increase market concentration, that it is intransparent for most users what kind of information about them is used for advertising purposes, and that users are not paid for the value creation they engage in when using commercial web 2.0 platforms and uploading data. Surveillance on Facebook is not only an interpersonal process, where users view data about other individuals that might benefit or harm the latter, it is economic surveillance, i.e. the collection, storage, assessment, and commodification of personal data, usage behaviour, and user-generated data for economic purposes. Facebook and other web 2.0 platforms are large advertising-based capital accumulation machines that achieve their economic aims by economic surveillance.

"The world will be better if you share more"? But a better world for whom is the real question? "Sharing" on Facebook in economic terms means primarily that Facebook "shares" information with advertising clients. And "sharing" is only the euphemism for selling and commodifying data. Facebook commodifies and trades user data and user behaviour data. Facebook does not make the world a better place, it makes the world a more commercialised place, a big shopping mall without exit. It makes the world only a better place for companies interested in advertising, not for users.

Facebook's understanding of privacy is property-oriented and individualistic. It reflects the dominant capitalistic concept of privacy. Needed are not only a socialist privacy concept and strategy, but an alternative to Facebook and the corporate Internet.

Facebook and Google are only the two most well-known examples for a more general contemporary economy that appropriates, expropriates, and exploits the common goods (communication, education, knowledge, care, welfare, nature, culture, technology, public transport, housing, etc.) created by humans and needed for all humans to survive. In the area of the Internet, a socialist strategy can try to resist the commodification of the Internet and the exploitation of users by trying to claim the common and participatory character of the Internet with the help of protests, legal measures, alternative projects based on the ideas of free access/content/software and creative commons, wage campaigns, unionisation of social media prosumers, boycotts, hacktivism, the creation of public service- and commons-based social media, etc. Internet exploitation is however a topic that is connected to the broader political economy of capitalism, which means that those who are critical of what social media companies like Facebook do with their data, ought better to be also critical of what contemporary capitalism is doing to humans throughout the world in different forms. If we manage to establish a participatory democracy, then a truly open society (Tännsjö 2010) might become possible that requires no surveillance and no protection from surveillance.

Notes

1 http://www.swissbanking.org/en/home/qa-090313.htm (accessed on 21 September 2010).
2 http://www.swissbanking.org/en/home/dossier-bankkundengeheimnis/dossier-bankkundengeheimnis-themen-geheimnis.htm (accessed on 21 September 2010).
3 Times (October 20, 2008, http://business.timesonline.co.uk/tol/business/industry_sectors/technology/article4974197.ece)

References

Andrejevic, Mark. 2002. The Work of Being Watched. *Critical Studies in Media Communication* 19 (2): 230–248.
Arendt, Hannah. 1958. *The Human Condition.* Chicago, IL: University of Chicago Press. Second edition.

Bellman, Steven, Eric J. Johnson, Stephen J. Kobrin, and Gerald L. Lohse. 2004. International Differences in Information Privacy Concerns: A Global Survey of Consumers. *The Information Society* 20 (5): 313–324.

Bennett, Colin and Charles Raab. 2006. *The Governance of Privacy.* Cambridge, MA: MIT Press.

Bolin, Göran. 2009. Symbolic Production and Value in Media Industries. *Journal of Cultural Economy* 2 (3): 345–361.

Bolin, Göran. 2005. Notes from Inside the Factory. *Social Semiotics* 15 (3): 289–306.

Bruns, Axel. 2008. *Blogs, Wikipedia, Second Life, and beyond. From Production to Produsage.* New York: Peter Lang.

Culnan, Mary J and Robert J. Bies. 2003. Consumer Privacy. Balancing Economic and Justice Considerations. *Journal of Social Issues* 59 (2): 323–342.

Dussel, Enrique. 2008. The Discovery of the Category of Surplus Value. In *Karl Marx's Grundrisse: Foundations of the Critique of the Political Economy 150 years later*, ed. Marcello Musto, 67–78. New York: Routledge.

Ess, Charles. 2009. *Digital Media Ethics.* Cambridge: Polity.

Etzioni, Amitai. 1999. *The Limits of Privacy.* New York: Basic Books.

Federal Trade Commission. 2000. *Privacy Online: Fair Information Practices in the Electronic Marketplace.* http://www.ftc.gov/reports/privacy2000/privacy2000.pdf (accessed on 26 March 2021).

Fuchs, Christian. 2011a. A Contribution to the Critique of the Political Economy of Google. *Fast Capitalism* 8 (1).

Fuchs, Christian. 2011b. *Foundations of Critical Media and Information Studies.* New York: Routledge.

Fuchs, Christian. 2010a. Labor in Informational Capitalism and on the Internet. *The Information Society* 26 (3): 179–196.

Fuchs, Christian. 2010b. Social Networking Sites and Complex Technology Assessment. *International Journal of E-Politics* 1 (3): 19–38.

Fuchs, Christian. 2010c. Some Reflections on Manuel Castells' Book "Communication power". *tripleC* 7 (1): 94–108.

Fuchs, Christian. 2010d. studiVZ: Social Networking Sites in the Surveillance Society. *Ethics and Information Technology* 12 (2): 171–185.

Fuchs, Christian. 2009a. Information Cnd communication Technologies and Society. A Contribution to the Critique of the Political Economy of the Internet. *European Journal of Communication* 24 (1): 69–87.

Fuchs, Christian. 2009b. *Social Networking Sites and the Surveillance Society. A Critical Case Study of the Usage of studiVZ, Facebook, and MySpace by Students in Salzburg in the Context of Electronic Surveillance.* Salzburg/Vienna: Research Group UTI.

Gandy, Oscar H. 1993. *The Panoptic Sort. A Political Economy of Personal Information.* Boulder, CO: Westview Press.

Habermas, Jürgen. 1989. *The Structural Transformation of the Public Sphere.* Cambridge, MA: MIT Press.

Hearn, Alison. 2010. Reality Television, the Hills, and the Limits of the Immaterial Labour Thesis. *tripleC* 8 (1): 60–76.

Held, David. 1996. *Models of Democracy.* Cambridge: Polity Press.

Hesmondhalgh, David. 2010. User-Generated Content, Free Labour and the Cultural Industries. *Ephemera* 10 (3/4): 267–284.

Jenkins, Henry. 2008. *Convergence Culture.* New York: New York University Press.

Jhally, Sut. 1987. *The Codes of Advertising.* New York: Routledge.

Jhally, Sut and Bill Livant, Bill. 1986. Watching as Working: The Valorization of Audience Consciousness. *Journal of Communication* 36 (3): 124–143.

Johnson, Deborah G. and Kent A. Wayland. 2010. Surveillance and Transparency as Sociotechnical Systems of Accountability. In *Surveillance and Democracy,* ed. Kevin D. Haggerty and Minas Samatas, 19–33. New York: Routledge.

Karatani, Kojin. 2005. *Transcritique: On Kant and Marx.* Cambridge, MA: MIT Press.

Kücklich, Julian. 2005. Precarious Playbour. *Fibreculture Journal* 5. http://five.fibreculturejournal.org/fcj-025-precarious-playbour-modders-and-the-digital-games-industry/ (accessed on 26 March 2021).

Lee, Micky. 2011. 2011. Google Ads and the Blindspot Debate. *Media, Culture & Society* 33 (3): 433–447.

Livant, Bill. 1979. The Audience Commodity. *Canadian Journal of Political and Social Theory* 3 (1): 91–106.

Lynd, Staughton. 1965. The New Radicals and "Participatory Democracy". *Dissent* 12 (3): 324–333.

Lyon, David. 1994. *The Electronic Eye. The Rise of Surveillance Society.* Cambridge: Polity.

Macpherson, Crawford Brough. 1973. *Democratic Theory: Essays in Retrieval.* Oxford: Clarendon Press.

Manzerolle, Vincent. 2010. Mobilizing the Audience Commodity: Digital Labour in a Wireless World. *Ephemera* 10 (3/4): 455–469.

Marcuse, Herbert. 1955. *Eros and Civilization.* Boston, MA: Beacon Press.

Marx, Karl. 1867. *Capital. Volume 1.* London: Penguin.

Marx, Karl. 1843a. Critique of Hegel's Doctrine of the State. In *Early Writings,* 57–198. London: Penguin.

Marx, Karl. 1843b. On the Jewish Question. In *Writings of the Young Marx on Philosophy and Society,* 216–248. Indianapolis, IN: Hackett.

Meehan, Eileen. 1993. Commodity Audience, Actual Audience. The Blindspot Debate. In *Illuminating the Blindspot,* ed. Janet Wasko, Vincent Mosco and Manjunath Pendakur, 378–397. Norwood, NJ: Ablex.

Mill, John Stuart. 1965. *Principles of Political Economy. 2 Volumes.* London: University of Toronto Press.

Moglen, Eben. 2003. *The dotCommunist Manifesto.* http://emoglen.law.columbia.edu/publications/dcm.html

Murdock, Graham. 1978. Blinfsports about Western Marxism. *Canadian Journal of Political and Social Theory* 2 (2): 109–119.

Negri, Antonio. 1991. *Marx beyond Marx.* London: Pluto.

Nissenbaum, Helen. 2010. *Privacy in Context.* Stanford, CA: Stanford University Press.

Pateman, Carole. 1970. *Participation and Democratic Theory.* Cambridge: Cambridge University Press.

Quinn, Michael. 2006. *Ethics for the Information Age.* Boston: Pearson.

Rule, James B. 2007. *Privacy in Peril.* Oxford: Oxford University Press.

Ryker, Randy, Elizabeth Lafleur, Chris Cox and Bruce Mcmanis. 2002. Online Privacy Policies: An Assessment of the Fortune E-50. *Journal of Computer Information Systems* 42 (4): 15–20.

Smythe, Dallas W. 1981/2006. On the Audience Commodity and its Work. In *Media and Cultural Studies*, ed. Meenakshi G. Durham and Douglas. M. Kellner, 230–256. Malden, MA: Blackwell.

Smythe, Dallas W. 1978. Rejoinder to Graham Murdock. Canadian *Journal of Political and Social Theory* 2 (2): 120–127.

Smythe, Dallas. 1977. Communications: Blindspot of Western Marxism. *Canadian Journal of Political and Social Theory* 1 (3): 1–27.

Starke-Meyerring, Doreen and Laura Gurak. 2007. Internet. In *Encyclopedia of Privacy*, ed. William G. Staples, 297–310. Westport, CT: Greenwood Press.

Tännsjö, Torbjörn. 2010. *Privatliv.* Lidingö: Fri Tanke.

Tavani, Herman T. 2008. Informational Privacy: Concepts, Theories, and Controversies. In *The Handbook of Information and Computer Ethics*, ed. Kenneth Einar Himma and Herman T. Tavani, 131–164. Hoboken, NJ: Wiley.

Toffler, Alvin. 1980. *The third wave.* New York: Bantam.

Zureik, Elia. 2010. Cross-Cultural Study of Surveillance and Privacy: Theoretical and Empirical Observations. In *Surveillance, Privacy and the Globalization of Personal Information*, ed. Elia Zureik, Lynda Harling Stalker, Emily Smith, David Lyon, and Yolane E. Chan, 348–359. Montreal: McGill-Queen's University Press.

References

Chapter Eight

Information Technology and Sustainability in the Information Society

8.1 Introduction

Sustainability has to do with the question of how present and future generations can lead a good life in society (for a review of its genesis, see Fuchs 2017). It is a concept that has been developed in forums such as the United Nations Conference on Environment and Development and the United Nations Conference on Sustainable Development. In the realm of information and communication technologies (ICTs), the sustainability concept has played a role in the context of the World Summit on the Information Society (WSIS). Sustainable ICTs have to do with the question if and how ICTs contribute and/or harm the development of society in ways that allow present and future generations to lead a good life.

This chapter asks: how can we think of sustainability and ICTs in the context of a critical theory of society? How is the sustainability of ICTs related to capitalism and class? The approach taken in this chapter stands in the tradition of critical sociology. This tradition "seeks to make problematic existing social relations in order to uncover the underlying structural explanations for those relations" (Fasenfest 2007, 17). Critical sociology is opposed to functionalism, is anti-positivist, uses the tradition of critical political economy, asks questions of power at large, and deconstructs ideologies (ibid.) Critical sociology is "a critique of the social order in the exploration of extant power relationships existing within a society organized under the principles of capitalist social relations" (Fasenfest 2007, 22). Its knowledge addresses "how to influence change toward a more progressive

DOI: 10.4324/9781003279488-10

and positive vision for the future" (Fasenfest 2007, 20). Given such a focus, it is evident that critical sociology is an approach suited for the study of (un)sustainable development.

Section 8.2 discusses the relationship of technology and capitalism. Section 8.3 identifies four ways of how to think of sustainability in the context of the information society. Section 8.4 criticises reductionist understandings of information technology sustainability. Section 8.5 provides a critique of dualistic understandings of information technology sustainability. Dualism and reductionism are the predominant mainstream concepts of sustainability in an ICT context. Section 8.6 suggests an alternative framework that uses critical theory as the foundation for a critical theory of sustainability in the information society.

8.2 Technology and Capitalism

The term technology has its roots in the Greek term techné [τέχνη] (Feenberg 2006; Reydon 2012; Williams 1983, 315), which means the knowledge, art, and craft of making something. Technology as techné was considered in subjective terms oriented on human skills, capacities, and knowledge to create something in a purposeful manner and thereby change the world. With the rise of modern large-scale industry and machinery, the dominant meaning of the category of technology shifted towards a more objective understanding. Technology has become to be understood as things, systems, machines, tools, artefacts, and hardware that apply the results of science for controlling humans and nature (see Li-Hua 2009; Williams 1983, 315).

Georg Lukács (1971, 131) argues that with the rise of capitalism, "human relations (viewed as the objects of social activity) assume increasingly the objective forms of the abstract elements of the conceptual systems of natural science and of the abstract substrata of the laws of nature". The economy thereby became "transformed into an abstract and mathematically orientated system of formal 'laws'" (105) that is governed by "the abstract, quantitative mode of calculability" (93). Technology in such a system is a machine that is used for controlling and instrumentalising nature and human activities for partial interests such as corporations' monetary profits and commodity production, bureaucratic power, possessive individualism, or consumerism.

Alfred Sohn-Rethel (1978) argues that this instrumental understanding of knowledge and technology goes back to the division of labour between manual and mental labour in class societies. The "logic of the market and of mechanistic thinking is a logic of intellectual labour divided from manual labour" (Sohn-Rethel 1978, 73). For Rethel, the logic of mechanistic, quantifying, and mathematical reasoning is not something that emerged with the existence of capitalism, but is much older. He argues that it goes as far back as ancient Greek slavery that instituted a division between manual labour performed by slaves and the mental labour of philosophers, politicians, and scientists. "It is Greek philosophy which constitutes the first historical manifestations of the separation of head and hand in this particular mode" (66). This division of labour has for Sohn-Rethel to do with the rise of the mathematical logic of measurement and quantification. Class society's division of labour would in the realm of thinking and logic be accompanied by quantifying reason and in the realm of the economy by exchange value.

In a general understanding, technology is neither knowledge nor a thing, but a process, in which humans make use of their skills, knowledge, and capacities and of objects in order to change the world in an intentional and purposeful manner. In modern class society, technology is no longer a human-controlled means for human-defined ends. Means and ends are reversed: humanity is not an end-in-itself, but humans have become means and instruments for dominant classes' partial interests (Fuchs 2016, Chapter 15). Technology is in this context an instrument for domination. Capital, including technology as its means of production, is a subject that dominates labour. Technology is in such a system not a means to humane ends, but rather servers a specific instrumental aim, namely capital accumulation, and as part of this end turns humans into objects.

The instrumental character of technology is not inherent in technology as such or in society in general, but rather has to do with how partial interests shape technology and society. Technology is not neutral and value free, but embedded into power structures, contradictions, and struggles that shape its invention, design, application, and use. This also means that technologies can be re-designed, re-invented, changed, re-purposed, abolished, etc. Putting technologies to humane and democratic use requires shaping society, invention, design, application, and use by humane and democratic values. It requires a political struggle for alternative technological and alternative frameworks that benefit all humans.

ICTs are means that humans use for creating, disseminating, and consuming information about the world. The computer and networked computer systems are particular technologies that other than traditional media (radio, television, the newspaper, etc.) do not just allow the consumption of information, but also its production, co-production, and dissemination.

The networked computer allows the convergence of the production, dissemination, and consumption of information in one tool. Given that technology is not independent of society, we cannot speak of the sustainability of technology just in technological terms, but need to connect this topic to society. A computer-controlled atom bomb is a particular political technology used for threatening actual or potential enemies. Its existence has to do with political power relations in the world. Defining technological sustainability immanently would mean that the atom bomb would be sustainable if it works error free, has comprehensive usability, can be controlled with the help of a user-friendly, secure and stable computer interface, etc. The problem with such an understanding is, however, that the computer-controlled atom bomb is inherently political and conflicts with the goal of a peaceful global society. It is politically unsustainable.

Such immanent definitions of technological sustainability that stay in the realm of technology without considering society often take on ideological forms. Mulder, Ferrer and van Lente (2011) argue that technological sustainability is not an end in itself: "Rather, sustainability of a technology can only be determined through a socio-political process" (Mulder, Ferrer and van Lente 2011, 242).

Computer technology cannot simply be made sustainable by changing chips, cables, variables, codes, or algorithms. Sustainable computing is not a technological matter because computing is embedded into environmental, economic, political and cultural contexts of design, production, and use. It is therefore necessary to discuss the topic of computing and sustainability in the context of the information society. Making computing sustainable requires shaping technology and society in an integrated manner (cp. Bijker, Hughes and Pinch 1987; MacKenzie and Wajcman 1999).

Such an understanding of technology is underlying the philosopher Ivan Illich's (1973) book *Tools for Conviviality*. He argues that it is dangerous to base society on what is technologically possible and not what is politically and ethically feasible. Illich argues that both society and technology need to be re-designed in an integrated manner. He therefore speaks of convivial tools in a convivial society. "Such a society, in which modern technologies serve politically interrelated individuals rather than managers, I will call 'convivial'. [...] I have chosen 'convivial' as a technical term to designate a modern society of responsibly limited tools" (Illich 1973, 12). We cannot assume that technological developments are automatically societally responsible.

Sustainability and technology development should be seen as two interlinked social and political tasks. One can certainly see the critique of unsustainable developments and technologies as a political task. At the same time, also reflecting on the implications of critical technology assessment for society and the construction of technology is an important political task. Schot and Rip (1996) argue in this context that constructive technology assessment and sociotechnical criticism are inherently connected.

For Illich, the problem is that technological innovations have the danger to blind people for potential negative consequences. Their all too optimistic adoption can backfire and result in unforeseen consequences. In an argument comparable to Illich, Horkheimer and Adorno (2002) argue that enlightenment reason can turn negatively against itself and have dangerous consequences. This is what they call the dialectic of the enlightenment. The implication of the problems that technologies can entail is to take an approach that tries to actively limit negative consequences by designing society and technology in human-centred ways. Such designs do not think primarily about what is "good for institutions" (Illich 1973, 25), but what is good for all humans.

8.3 Four Approaches to Understanding Sustainability in the Information Society

Discussions about the un/sustainability of information technology's role in society have especially emerged since the First WSIS that was held in two phases in 2003 and 2005.

We can classify information society policy discourses according to how they relate the domains of the ecology and the economy to the realms of politics and culture. According to the information philosopher Wolfgang Hofkirchner (2013), there are four ways of how the relationship of two categories C1 and C2 can be explained: reductionism, projectivism, disjunctivism/dualism, and dialectical integrativism. Reductionism causally reduces the relation C1–C2 to C1. Projectivism projects causality into C2. Dualistic thought argues that C1 and C2 have independent causalities. A dialectical approach sees C1 and C2 as at the same time relatively autonomous and mutually constituting each other. In a dialectic, C1 and C2 are identical and non-identical at the same time. I have in other publications elaborated and applied based on Hofkirchner's typology a distinction of four information society policy discourses (Fuchs 2010; Fuchs and Verdegem 2013) (Table 8.1).

Reductionist approaches see ecological or technological or economic developments (such as GDP investment in information technology and the information economy) as the sole driving forces of the un/sustainable information society. Projectivist approaches see the political and/or cultural system as the determining forces of un/sustainability in the information age. Reductionism sees the physical aspects of society as determining, whereas projectivism assumes that the realms of human ideas and politics are determining society's development. Dualistic approaches define multiple goals and dimensions

TABLE 8.1   Approaches on sustainability and information society policies (based on: Fuchs 2010)

Type of approach	Description
Reductionism	Ecology, economy, or technology are considered as the driving forces of a sustainable information society.
Projectivism	Politics and/or culture are seen as the determinant forces of a sustainable information society.
Dualism	Multiple dimensions and goals of a sustainable information society are identified, but not causally related to each other.
Dialectic	Multiple dialectically interrelated dimensions and goals of a sustainable information society are identified, existing contradictions of these dimensions are analysed, and changes are seen as integral, interdependent, and systemic.

of a(n) un/sustainable information society, but do not consider if these goals are compatible and if and how they are causally linked. Dialectical approaches see the various dimensions and goals of un/sustainability in the information society as interdependent, mutually causally linked, and only relatively autonomous.

Projectivism is an approach that can hardly be found in ICT policy discourses on sustainability because the notion of sustainability originates in the environmental realm and this kind of discourse tends to be associated with industry interests. Therefore either the ecological or the economic or both dimensions normally tend to play a role. Theoretically, ICT sustainability could of course be conceived in purely political or cultural terms with a pure focus on either digital democracy or fostering online understanding. Reductionist understandings are much more common than projectionist ones.

8.4 Reductionist Understandings of Sustainability in the Information Society

Hilty and Ruddy (2010) reject multidimensional definitions of sustainability in general and in the ICT context in particular because they argue that nature is the most fundamental dimension of human survival. They say that "multidimensionality mitigates the radical nature of SD" (Hilty and Ruddy 2010, 11). They define the central concern of sustainable development as the "sustainability dilemma", i.e. "the physical impossibility of extending the present consumption patterns of the industrialized countries to all parts of the world without putting a great burden on future generations" (Hilty and Ruddy 2010, 10) and reduce the sustainability of IT to the ecological dimension.

The emergence of ICTs and the Internet has not dematerialised the economy. The depletion of non-renewable natural resources and the massive emission of carbon dioxide continue. The ecological catastrophe is certainly an important challenge in the information society. But assume that we had solved this problem, then other ways of destroying humanity could nonetheless still persist, especially politically and ideologically motivated wars and spirals of violence that in escalation could result in the large-scale use of nuclear, chemical, and biological weapons that could wipe out humanity. Also, economic crises have the potential to render the lives of many people precarious and can lead to political crises and in the last instance also to wars. The example of dematerialisation's promises shows that in a society, where groups compete for resources (including capital, influence, attention, support, etc.), technological determinism is used as a means in the struggle for mobilising resources for political interests.

Hilty and Ruddy create the impression that the environmental crisis is the only problem that needs to be solved in the information age. Their approach is a form of **environmental reductionism**. We can also not exclude the possibility that it may indeed be possible to universalise today's per capita quantity of physical consumption to all humans if it is at the same time possible to make a large-scale qualitative shift to green energy and renewable resource use. Given that there is more than one dimension that threatens the existence of humanity and the attainability of a good society, a one-dimensional use of the category of the sustainable information society is not feasible.

The European Union in 2010 introduced its new information society policy called *A Digital Agenda for Europe*, in which it formulates a policy strategy and goals it wants to reach until 2020 (European Commission 2010). "The overall aim of the Digital Agenda is to deliver sustainable economic and social benefits from a digital single market based on fast and ultra-fast Internet and interoperable applications" (European Commission 2010, 3). The notion of sustainability is here used as both meaning (a) the continuous growth of profits and the GDP as well (b) the continuous guarantee of social cohesion. There is no consideration that there may be an antagonism between on the one side the focus on companies' profits, which in the past decades has in most part of the world meant a neoliberal policy agenda, and on the other side the increasing social inequalities that have come along with neoliberalism. The overall aim formulated in the Digital Agenda is both **economic reductionist and technologically deterministic:** The EU assumes that the combination of the Internet and neoliberalism automatically brings about economic and social sustainability.

The EU expresses its view that the Internet in Europe is not developed enough, not fast enough and that the uptake is not widely enough:

> "More needs to be done to ensure the roll-out and take-up of broadband for all, at increasing speeds, through both fixed and wireless technologies, and to

Reductionist Understandings of Sustainability in the Information Society

facilitate investment in the new very fast open and competitive internet networks that will be the arteries of a future economy. Our action needs to be
focused on providing the right incentives to stimulate private investment, complemented by carefully targeted public investments, without re-monopolising
our networks, as well as improving spectrum allocation".

(European Commission 2010, 6)

One of the keywords of the EU for creating sustainability is the focus on a "vibrant digital
single market" (7) for Internet services, digital content and "telecom services" (7), which
includes Internet access and infrastructure.

*"We need very fast Internet for the economy to grow strongly and to create jobs
and prosperity, and to ensure citizens can access the content and services they
want.* The future economy will be a network-based knowledge economy with
the internet at its centre. Europe needs widely available and competitively-
priced fast and ultra fast internet access. The Europe 2020 Strategy has under-
lined the importance of broadband deployment to promote social inclusion and
competitiveness in the EU".

(18–19)

The EU has the objective to achieve "broadband for all" (26) and wants to specifically
foster the deployment of Next Generation Access (NGA) networks (20), which are Internet networks that have a download speed of more than 24 Mbit/s. The EU strategy in
this respect is to "encourage market investment in open and competitive networks" (20).

The EU overall fosters a neoliberal approach to digital society's sustainability. There are
of course exception, such as the EU research project netCommons (http://www.netcom-
mons.eu) that stresses that we need alternative technological, legal, political, social,
ethical, and economic frameworks for advancing the sustainability of the information
society. The EU sees capitalist businesses as the key to providing Internet access and
services and sees Internet capitalism as the source of the growth of economic profitability and the creation of wealth and social inclusion. In the above-mentioned example
quotes, social goals are reduced to an economic dimension, namely, the advancement of
digital capitalism. The Digital Agenda overlooks that capitalist investments in Internet
access and services do not guarantee social cohesion. Capital has the inherent drive to
increase itself and as one of its means for accumulations tends to aim at a reduction
of wage costs. Precarious and unpaid digital labour, i.e. labour that produces digital
media technologies and services, has been one of the effects of the capitalist Internet
economy (Fuchs 2014b).

Regional development is an aspect of sustainable development. If certain regions are significantly worse off than others, then regional inequality constitutes a form of unsustainability. The EU considers regions that have a per capita GDP lower than 75% of the EU average as being less developed. In the years 2014–2020, this includes all of Bulgaria, Estonia, Latvia, Lithuania, Romania, Slovakia, Slovenia as well as parts of Croatia, the Czech Republic, Greece, Hungary, Poland, Portugal, South Italy, Spain, and the UK (Cornwall, West Wales). These regions are especially located in Europe's South and East, which is an indication of uneven development in Europe (data source: http://ec.europa.eu/regional_policy), accessed on 12 December 2020). Table 8.2 shows that in the EU, less developed regions, sparsely populated areas, poor households, and individuals with low education have significantly lower use of the Internet and computers than the average EU citizen.

There is less Internet use in less developed regions (data source: Eurostat). Tables 8.3 and 8.4 show the regions in Europe that in 2015 had the largest share of citizens who had never used a computer and the lowest use of broadband Internet. They again indicate that it is regions in the South and East of Europe that have the lowest computer and Internet use.

TABLE 8.2 Internet and computer use statistics for the EU (data source: Eurostat)

	Individuals regularly (at least once a week) using the Internet, 2015	Households with Internet access, 2015	Households with broadband access, 2015	Households owning a computer, 2015	Share of individuals who have never used the Internet, 2015
EU28	76%, 2010: 65%	83%	80%	82%	16%
Less developed regions	2013: 59% total EU28 (2013): 72%	2013: 68%, total EU28 (2013): 79%	2013: 66%, total EU28 (2013): 76%	2013: 70%, total EU28 (2013): 80%	2013: 31%, total EU28 (2013): 20%
ICT professionals	92%				3%
Manual workers	72%				17%
Low education	55%				36%
Individuals in poorest households (lowest quartile)	48%	62%	59%	62%	31%
Individuals in richest households (upper quartile)	81%	97%	95%	97%	5%
Households in sparsely populated areas (<100 inhabitants/km^2)			73%	77%	23%

TABLE 8.3 Regions in the EU, where in 2015 less than 60% of households had broadband access at home (data source: Eurostat)

Region	%
Severozapaden, Bulgaria	45
North and South Bulgaria	55
Severoiztochen, Bulgaria	56
Yuzhen tsentralen, Bulgaria	56
Corsica, France	57
Macroregiunea, Romania	57
Nord-Est, Romania	57
Sud-Est, Romania	57
Severen tsentralen, Bulgaria	58
Yugoiztochen, Bulgaria	58
Central Greece	59

TABLE 8.4 Regions in the EU, where in 2015 40% or more have never used a computer (data source: Eurostat)

Region	%
Severozapaden, Bulgaria	49
Campania, Italy	42
Apulia, Italy	42
Sud, Romania	40
Molise, Italy	40
Sicily, Italy	40

Given the existence of a digital divide between poor citizens and regions on the one side and rich citizens and regions on the other side in Europe, the question arises if an approach that fosters private ownership and for-profit operation of Internet networks is suited for overcoming such divides. For-profit means that operators charge for network access. Access is organised as a commodity. Given income inequality, those on lower income are less likely to afford the same level and speed of access than those who are better off. Capitalist markets necessarily bring access inequalities with them.

The EU, however, follows predominantly a market approach in the creation of fast broadband networks. In 2014, the EU announced the European Fund for Strategic Investments (EFSI), a plan of investing 315 billion Euros into broadband infrastructure, transport, education, research, and innovation in the years 2015–2017 as a combination of public funding and private investment.[1] Around 80% comes from private investors, the rest from the European Investment Bank and the European Investment Fund (European Commission 2016).

> "The Investment Plan for Europe adopted in November 2014 as the first major initiative of the Juncker Commission has the potential to bring investments back in line with its historical trends. Via the EFSI, the European Investment Bank is able to respond quickly to financing needs in areas where alternative sources of financing are scarce or unavailable. The Bank's presence often provides reassurance to other financiers to provide co-financing. The EFSI projects need to be economically and technically viable, consistent with Union policies, provide additionality (i.e. they could not be realized without the backing of the EU guarantee), and maximise the mobilisation of private sector capital".
>
> (European Commission 2016)

The President of the European Commission Jean-Claude Juncker commented:

> "We need to pursue fiscal responsibility and keep public finances sustainable. We also need to restore investment levels to overcome the crisis, to kick-start growth and sustain it. [...] We have to [...] stimulate private capital. We cannot spend money we do not have. So this is an offer to the private sector where the money is [...] to join the efforts we are developing".[2]

The discussion shows that there is a policy regime in Europe that tends to foster Internet infrastructure and access as a commodity. There is not just unequal access to the Internet in Europe, but also a large market concentration in the broadband market. Since 2012, over 60 billion Euros were spent on mergers and acquisitions of telecommunications operators in the EU (European Commission 2015). In 8 of 28 EU countries, the incumbent controls more than 50% of all broadband subscribers (ibid.): Luxemburg, Cyprus, Austria, Denmark, Estonia, Latvia, Croatia, and Lithuania. For all of Europe, incumbents control 41% of the subscribers (ibid.). Table 8.5 provides an overview of the dominant market player's share in broadband subscriptions for all European countries.

The Herfindahl–Hirschman Index (HHI) is a measure of market concentration. It is calculated the following way:

$$HHI_j = \sum_{i=1}^{f} S_{ij}^2$$

f = number of firms participating in an industry,

TABLE 8.5 Market share of the incumbent in fixed line broadband subscriptions and minimum level of the Herfindahl–Hirschman Index, data for 2015, data source: European Commission 2015

Country	Share (%)	HHI >
Luxembourg	69	4,761
Cyprus	64	4,096
Austria	58	3,364
Denmark	58	3,364
Estonia	58	3,364
Latvia	58	3,364
Croatia	53	2,809
Lithuania	51	2,601
Malta	49	2,401
Portugal	48	2,304
Italy	48	2,304
Spain	45	2,025
Belgium	44	1,936
Hungary	44	1,936
Greece	43	1,849
Germany	42	1,764
Netherlands	41	1,681
France	39	1,521
Sweden	39	1,521
Ireland	37	1,369
Slovenia	35	1,225
Slovakia	34	1,156
UK	32	1,024
Poland	32	1,024
Czech Republic	29	841
Romania	27	729
Bulgaria	23	529
Average in EU	44	HHI > 2,106

S_{ij} = each firm i's market share in the industry j,

HHI < 1000: low market concentration,

1000 < HHI < 1800: moderate market concentration,

HHI > 1800: high market concentration (Noam 2009, 41).

The calculations of the HHI in Table 8.6 show minimum levels. We can infer from them that in at least 15 of 27 EU countries, for which data is available, the broadband market was highly concentrated in 2015. The average EU HHI in the broadband market is at least 2106, which is also a very high level.

Mobile broadband has a relatively small share of the broadband market: In 2014, only 8.3% of the homes in the EU used mobile Internet connections for accessing the Internet (European Commission 2015, 29). Table 8.6 shows that the average minimum HHI for the

TABLE 8.6   Market share of the incumbent in mobile network subscriptions and minimum level of the Herfindahl–Hirschman-Index, data for 2014, data source: Eurostat (Digital Agenda Key Indicators), UK and Germany: Ofcom (2015)

Country	%	HHI >
Cyprus	66	4,338
Luxembourg	55	2,973
Slovenia	48	2,345
Portugal	47	2,246
Croatia	46	2,146
Hungary	45	2,049
Malta	44	1,968
Romania	44	1,933
Lithuania	43	1,815
Austria	42	1,776
Latvia	42	1,769
Slovakia	42	1,734
Estonia	41	1,665
Finland	40	1,587
Czech Republic	39	1,556
Denmark	39	1,524
Ireland	38	1,439
Bulgaria	37	1,369
Germany	37	1,369
Sweden	36	1,299
France	33	1,106
Spain	32	1,025
Italy	32	996
UK	30	900
Poland	30	888
EU	41	1,753

Reductionist Understandings of Sustainability in the Information Society

mobile communications market in 25 EU countries in the year 2014 was 1,753. Given that this is a minimum value based on the market share of only the incumbent, we can assume that the actual value is higher than 1,800 and that therefore also the European mobile communications market is highly concentrated.

Strong market concentration means that economic power is asymmetrically distributed. Single companies have economic advantages at the expense of workers in other companies in the same sector, whose economic survival is threatened. Market concentration also enables price control. An economy characterised by corporate monopolies or oligopolies can therefore not be considered as being sustainable. In the EU, there is the dominance of a profit-oriented telecommunications and Internet model, in which large telecommunications corporations have lots of power. Given that the capitalist development of the Internet market has resulted in high broadband market concentration, the question arises if it is wise to further foster the market model in building new infrastructure or if alternative models are needed. The EU's strategy to try to stimulate private investments into Internet infrastructure can easily enforce further market concentration: investments into communications infrastructure are very expensive because it involves the digging of trenches and the laying of fibre cables and ducts. Only companies with lots of capital can undertake such investments. Given a high concentration of communications markets as in Europe, the most likely investors into new communications infrastructure are the incumbent players, which strengthens their market advantages, makes it more likely that they also dominate the new markets, which then reinforces capital concentration.

The EU example shows that *fostering private investments with the help of public aid in an overall highly concentrated economic realm such as communications tends to reinforce concentration.* We can therefore speak of a vicious cycle of capital concentration in the communications infrastructure market. Furthermore, communications corporations such as Verizon, Vodafone, EE, O2,[3] Hutchison, Tele Columbus, Tele2, and Telecom Italia[4] seem to have avoided paying taxes in Europe. The argument that private investment is needed because public finances are under strain seems to overlook that public funding could certainly be increased if tax avoidance structures could be overcome and large corporations be made accountable.

Such processes constitute together a *vicious cycle of neoliberalism* that operates in the communications market and other markets (see Figure 8.1): Neoliberal policies and ideology foster the commodification of services, society's resources, infrastructures, and services (Harvey 2005). The result is the emergence of capitalist markets. Markets in

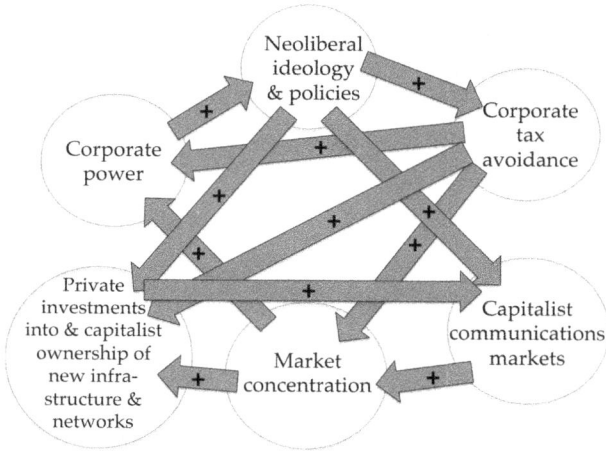

FIGURE 8.1   The vicious cycle of neoliberalism.

general have a tendency to concentrate and form oligopolistic and monopolistic structures. Communications markets are affected by concentration in a particular way: investment into network infrastructures and information technologies are expensive, which fosters concentration. Advertising-funded media tend to attract advertisers if they attract large numbers of viewers, readers, listeners, and users, which fosters the concentration of advertising via an advertising-audience share-spiral (Furhoff 1973). Selling media content is a high-risk business, in which survival is difficult. All of these mechanisms foster the concentration of communications markets. Neoliberalism also fosters a tendency for corporate tax avoidance that together with concentration tendencies strengthens the power of corporations. Building, maintaining, and operating communications infrastructure are expensive. Given market concentration, especially existing incumbent operators tend to be able to afford necessary investments so that there is a tendency that dominant market actors tend also to control new communications infrastructures. Corporate tax avoidance not just strengthens the financial power of corporations, but also puts pressure on public finances to further foster neoliberal policy agendas. Increasing corporate power fosters the tendency that corporations are enabled to threaten state institutions to withdraw or outsource their capital, which may result in unemployment. The neoliberal competition state competes with other states for attracting capital and so tends to foster ever more commodification, privatisation, and market liberalisation. The outcome is a vicious cycle of neoliberalism, in which neoliberal policy and ideology, capitalist markets, market concentration, and corporate power are reinforced. The example analysis shows that such a vicious cycle operates in the European Union's information society.

Reductionist Understandings of Sustainability in the Information Society

Overall the example of broadband markets in Europe's confirms the analysis that the EU's Digital Agenda is based on a neoliberal economic reductionism that fosters the market and capitalism in the realm of digital media and sees the market as primary force for sustainability.

(for a critique of neoliberalism, see also: Harvey 2005; Saad-Filho and Johnston 2005)

8.5 Dualistic Understandings of Sustainability in the Information Society

The **third type of information society policy discourse** is **dualistic** in character. The World Summit of the Information Society was a summit that the United Nations organised. It took place in two phases, with one event in 2003 in Geneva and another one in 2005 in Tunis.

WSIS identified the potentials of ICTs to eradicate hunger and poverty and foster education, gender equality, health care, environmental sustainability, peace, prosperity, freedom, democracy, human understanding, cultural diversity, and human rights (WSIS 2003, §§2, 3, 51). It argued that GDP growth and social equality can be advanced at the same time through ICTs: "Under favourable conditions, these technologies can be a powerful instrument, increasing productivity, generating economic growth, job creation and employability and improving the quality of life of all. They can also promote dialogue among people, nations and civilizations" (WSIS 2003, §9).

WSIS' logic of argumentation is dualistic because it assumes that through ICT development both capitalist growth and social equality can be achieved at the same time. ICT development is seen as a realm of capitalist investment, both in developed and developing countries: WSIS promoted ICT and Internet development in developing countries through the support of foreign direct investment and transfer of information technology (WSIS 2003, §40; see also WSIS 2005, §§54+90b). It encouraged "private-sector participation" (WSIS 2005, §13) and identified a "powerful commercial basis for ICT infrastructural investment" in developing countries (WSIS 2005, §14). It wanted to "promote and foster entrepreneurship" in the realm of ICTs in developing countries (WSIS 2005, §90b) and spoke of "sustainable private-sector investment in infrastructure" (WSIS 2005, §20). We can here find a peculiar understanding of sustainability as "private-sector investment in infrastructure". Sustainability is here not related to the common good that benefits all, but to the growth of the profits of private companies that own Internet infrastructure. In a comparative passage, WSIS called for "adequate and sustainable investments in ICT infrastructure and services" (WSIS 2005, §8). WSIS calls for both private ownership and social benefits for all at the same time.

In contrast to WSIS, the winners of the Noble Prize in Economics Joseph Stiglitz (winner in 2001) and Amartya Sen (winner in 1998) argue that capitalist growth is no guarantee for social justice as aspect of sustainability. Stiglitz, Sen and Fitoussi (2010) write that the GDP is of limited use for measuring social progress and that it is "an inadequate metric to gauge well-being over time" (Stiglitz, Sen and Fitoussi 2010, 8). Measuring well-being by the GDP could for example "send the aberrant message that a natural catastrophe is a blessing for the economy, because of the additional economic activity generated by repairs" (Stiglitz, Sen and Fitoussi 2010, 265). They call for a shift in emphasis "*from measuring economic production to measuring people's well-being*" (12) in policymaking and research in the context of sustainability.

The WSIS meetings in 2003 and 2005 were based on a neoliberal policy agenda that advances a dualistic agenda that sees social sustainability and capitalist growth of profits as achievable by capitalist ownership and the development of ICT infrastructure. Since the rise of neoliberal politics that advanced privatisation, the commodification of common goods and public services, market liberalisation, and the deregulation of social policies, inequality understood as the distribution of income between labour and capital and between the rich and the poor, resource inequality, and inequality of health and death has increased.

> "In March 2008, before the bubble burst, *Forbes* magazine listed 1,125 of the world's billionaires. Together, they owned $4.4 trillion. That was almost the entire national income of 128 million Japanese or a third of the income of 302 million Americans".
>
> (Therborn 2012, 584)

WSIS propagated a so-called "multi-stakeholder approach" that in Internet governance fosters the co-operation of "governments, the private sector, civil society and other stakeholders, including the international financial institutions" (WSIS 2003, §60; see also WSIS 2005, §§29, 34, 80, 83, 97, 98). Such formulations create the impression that these actors possess equal shares of power in the world. Transnational corporations have significant shares of money, reputation, and influence and may therefore be more capable of being more heard in policy debates and policy formulations than civil society actors. It is therefore not a surprise that in contrast to the official "multi-stakeholder" documents published by WSIS in 2003 and 2005 that have a corporate-friendly character, the 2003 Civil Society Declaration to the WSIS formulated a different vision. It said that "full participation in information and communication societies requires us to reject at a fundamental level, the solely profit-motivated and market-propelled promotion of ICTs for development. Conscious and purposeful actions need to be taken in order to ensure that new ICTs are not deployed to further perpetuate existing negative trends of economic globalisation and market monopolisation".

(WSIS Civil Society Plenary 2003, 7)

WSIS saw public service investment and provision of Internet access only feasible in poor regions: "We recognize that public finance plays a crucial role in providing ICT access and services to rural areas and disadvantaged populations including those in Small Island Developing States and Landlocked Developing Countries" (WSIS 2005, §21). It did not consider that capitalist ownership of communications infrastructure tends to be, as we have already seen, economically highly concentrated, which means a high concentration of power and private wealth. Public service infrastructure in a world of high inequality and concentrated capitalist ownership may therefore be a feasible alternative not just for developing regions. The argument that the public should only step in where private investors cannot easily make profits overlooks that the market also fails in other areas, where transnational corporations make large profits and such accumulation results in market concentration.

Ten years after the WSIS, the WSIS+10 High Level Event conducted a progress review (Geneva, 10–13 June 2014) and published outcome documents. The approach has ten years later not changed and remains dualist: ICTs are "cross-cutting enablers for achieving the three pillars of sustainable development" (WSIS+10 2014, 10). WSIS+10 recognises some problems such as the gender digital divide, the lack of youth empowerment, the lack of Internet access in the least developed countries, that the voluntary digital solidarity fund does not work, e-waste, or privacy issues resulting from mass surveillance. But overall it is just like the WSIS outcome documents in 2003 and 2005 over-confident that capitalism and the market are the right way to social and economic progress.

The WSIS agenda is still dualist: "ICTs should be fully recognized as tools empowering people, and providing economic growth" (12). And it is also still neoliberal, although the new world economic crisis has shed doubts on this approach. "To attract private investment, competition and adequate market liberalization policies to develop the infrastructure, financing, and new business models need to be studied and deployed, taking into account national circumstances" (WSIS+10 High Level Event 2014, 36). "We recognize the critical importance of private sector investment in information and communications technology infrastructure, content and services, and we encourage Governments to create legal and regulatory frameworks conducive to increased investment and innova-tion" (United Nations General Assembly 2015, §38).

WSIS simply ignores certain important issues that concern the development of the in-formation society and show the latter's contradictions in capitalism: the concentrated wealth of the rich (including the owners and CEOs of the largest transnational com-munications corporations), precarious labour (especially in the younger generation), computerisation- and automation-induced unemployment, the crisis of capitalism, profit/wage-inequality, income and wealth inequality, the concentration of ownership in the communications industries, unpaid and precarious digital and crowdsourced labour, com-munications corporations' tax avoidance, etc.

In 2015, there were 241 information companies among the world's 2,000 largest transna-tional companies.[5] Together they had combined profits of US$537.3 billion (data source: Forbes 2000, 2015 list). These profits exceeded the combined GDP of the world's 33 least developing countries (US$474.0 billion) and the combined GDP of the world's 74 smallest economies (GDP of US$536.2 billion) (data source: UNHDR 2015, World Bank Data [GDP I market prices in current US$])). Table 8.7 shows the world's ten most profitable transna-tional information corporations in the year 2015.

The combined profits of the world's ten largest transnational information corporations (US$240.0 billion) are larger than the combined GDP of the world's 16 least developed countries (US$229.2 billion) and larger than the combined GDP of the world's 54 smallest

TABLE 8.7   The world's most profitable transnational information corporations in the year 2015 (data source: Forbes 2000, 2015 list)

No.	Forbes rank	Company	Industry	Profits 2015 (billion US$)
1	40	Vodafone	Telecommunications	77.4
2	12	Apple	Computer hardware	44.500
3	18	Samsung Electronics	Semiconductors	21.9
4	25	Microsoft	Software and programming	20.7
5	20	China Mobile	Telecommunications	17.7
6	39	Google	Computer services	13.700
7	44	IBM	Computer services	12.000
8	67	Intel	Semiconductors	11.7
9	88	Oracle	Software and programming	10.8
10	22	Verizon	Telecommunications	9.6
				Total: US$240.0 bn

economies (US$234.2 billion) (data source: UNHDR 2015, World Bank Data [GDP at market prices in current US$]). Vodafone was in 2015 the world's most profitable transnational information corporation. Its profits amounted to US$77.4 billion. Vodafone's profits were larger than the individual economic performance of 114 of the world's countries (data source: World Bank Data, GDP at market prices in current US$ for 2015), including populous countries such as Ethiopia (100 million inhabitants), the Democratic Republic of Congo (75 mn), Tanzania (52 mn), Kenya (45 mn), Uganda (38 mn) (data source: World Bank Statistics, year 2014). Vodafone, a British telecommunications company that uses "a Luxembourg entity to reduce tax bills", according to reports paid no corporation tax in 2014/2015.[6]

These data show the power of transnational information corporations. They are very profitable companies. Their individual economic power is often larger than the one of entire countries. Their profitability is often further increased by tax avoidance. At the same time, there is large inequality between profits and wages, and neoliberalism and austerity measures have resulted in cuts of social expenditures and the rollback and privatisation of public services. Talking about the sustainability of the information society without talking about the profits of information corporations and the wealth of the rich, as the WSIS does, has a quite ideological character. Dualistic thought formulates the goal of corporate profitability together with a wish list of social equality goals and ignores the actual contradiction between the first and the second. Can there be an alternative, critical understanding of sustainability in an information society and information technology context?

8.6. Conclusion: Towards a Critical, Dialectical Understanding of Sustainability in the Information Society

This chapter has shown that there are two dominant versions of ICT sustainability discourse: neoliberal reductionism and neoliberal dualism. They have in common that they both have a quite similar role in society as ideologies that try to legitimatise the dominant way of how corporations and politicians organise information technologies as instruments for corporate and bureaucratic control.

Habermas (1968/1989) stresses in this context based on Herbert Marcuse (1964) that technology becomes a form of technological rationality that is a form of domination and ideology. The analysis presented in this chapter shows that Marcuse's and Habermas's insights about technological rationality remain highly relevant in the time of social computing, the Internet, cloud computing, the Internet of things, and big data. Reductionist and dualist versions of ICT sustainability are one-dimensional and instrumental concepts of the relationship between information technology and society. They are means for domination and ideological legitimation. Langdon Winner (1986, 105) speaks of mythinformation as the "conviction that a widespread adoption of computers and communication systems along with easy access to electronic information will automatically produce a better world for human living". Mythinformation is the ideology of those "who build, maintain, operate, improve, and market" as well as regulate and analyse computing systems in an instrumental manner (Winner 1986, 113). Reductionist and dualist versions of ICT sustainability believe in specific versions of mythinformation, namely, that the combination of computing and capitalism will produce a better world.

Winner (1986) describes how in respect to computing, political questions such as "How can we live gracefully and with justice?" (162), "Are we going to design and build circumstances that enlarge possibilities for growth in human freedom, sociability, intelligence, creativity, and self-government? Or are we headed in an altogether different direction?" (17), or "How can we limit modern technology to match our best sense of who we are and the kind of world we would like to build?" (xi) are often simply not asked. Today, the moral values of computing are being discussed in the context of buzzwords such as sustainability and corporate social responsibility. Corporations, managers, bureaucrats, and instrumentalists can no longer simply ignore moral philosophy. Today moral questions tend to be asked, but the answers remain one-dimensional and naïve, often stressing that "Information technology will fix society's problems" (technological reductionism), or "Capitalist adoption of information technology will fix society's problems" (economic reductionism), or "We want to have capitalism, capitalist information technologies and a good society" (dualism).

Critical theorists of technology and society share the insight that alternative models of technology and society that transcend instrumental reason are needed. Marcuse speaks in this context of the need for dialectical rationality (Marcuse 1964) and technologies of liberation (Marcuse 1969), Habermas (1968/1989) of communicative action, Illich (1973) of convivial tools, and Raymond Williams (1976) of democratic communications. Winner (1986, Chapters 4 and 5) reminds us in his critique of decentralised, appropriate technologies that there is no alternative technology-fix and no alternative consumer culture-fix to society's problems. "Appropriate technologists were unwilling to face squarely the facts of organized social and political power" (Winner 1986, 80). The implications for alternative computing, networking, online and Internet technologies today are that centralised power exists in a technologically decentralised world and that alternative digital technologies not just require alternative designs that foster democratic alternatives, but also struggles for the democratisation of the institutions, contexts and society, in which alt-tech is used. The struggle for alternative technologies must at the same time be the struggle for an alternative society, a participatory democracy.

A **dialectical perspective on the information society** is based on these insights and sees unsustainable development as the result of contradictions in society that are mediated by information technology and result in destruction and inequalities. Table 8.8 gives an overview of the dimensions of un/sustainable ICTs and an un/sustainable information society.

What do we mean if we speak of a dialectic (for a more detailed discussion, see: Fuchs 2014a, 2011, Chapter 2.4)? A dialectic is a contradictory relationship between two entities. They simultaneously are identical and different. They require and exclude each other. Dialectical logic challenges classical binary and reductionist thought. It questions the reduction of the world to just one dimension. It is, however, not just relational and multidimensional, but also sees the world as being in flux and development. Development potentialities emerge out of poles that contradict each other. At a certain level of organisation, everything constantly develops. There are, however, also more continuous processes that only change at specific critical points. Dialectical development includes situations of crisis and change and the emergence of novelty at such critical points. In society, there are two basic forms of the dialectic: One has to do with the very basic conditions and the basic development of society. So for example there is a social dialectic between human beings: In order to exist, humans have to communicate with each other. They are different individuals, but can only inform themselves by mutual symbolic interaction. The second form of societal dialectic has to do with power relations. In a power dialectic, we find conflicting interests and conflicting structures.

The basic assumption, on which a dialectical concept of un/sustainable ICTs in an un/sustainable information society is based, is that unsustainability means that there are contradictory interests in the production and/or use of digital media technologies, such as for example a contradiction between nature and society (environmental unsustainability), digital capital and digital labour (economic unsustainability), the rulers and the ruled (political unsustainability), or a cultural elite and everyday people (cultural unsustainability).

TABLE 8.8 A dialectical view of the un/sustainability of ICTs and the information society

Dimension	Dimension of sustainability	Question	Dimension of unsustainability	Question
Nature	Environmental sustainability of ICTs: Biodiversity (Questions concerning eWaste and the energy consumption of ICTs)	To which degree does ICT use respect the protection and preservation of natural resources so that the survival of nature and society is guaranteed? To which degree is there an equitable distribution of ICTs' environmental harms and benefits to certain groups and places?	Environmental unsustainability of ICTs: Contradiction between nature and society (environmental pollution, degradation and depletion)	To which degree does ICT use result in the depletion of non-renewable natural resources, the consumption of non-renewable energy resources, the production of non-recyclable (e-)waste, and in pollution? To which degree is there an unequal and inequitable distribution of ICTs' environmental harms and benefits to certain groups and places?
Society: Economy	Economic sustainability of ICTs: Wealth for all (Questions concerning power, monopolies, labour, access, affordability, and resource availability in the digital media industry)	To which degree is a social system that produces, uses or provides access to ICTs organised in a way that fosters wealth for all and a fair distribution of wealth?	Economic unsustainability of ICTs: Contradiction between digital capital and digital labour (poverty, inequality, economic crisis)	To which degree is a social system that produces, uses or provides access to ICTs organised in a manner that does not guarantee the satisfaction of the needs of all humans (poverty), that results in unfair distribution of need satisfaction (inequality) or the irreproducibility of the economy (economic crisis)?
Society: Political system	Political sustainability of ICTs: Participation and peace (Questions about eParticipation, eDemocracy, cyberwar, online privacy, digital surveillance)	To which degree does the social organisation underlying the production or use of ICTs enable humans to participate in collective decision-making? To which degree does the use of ICTs guarantee the peaceful existence and interaction of societies and the guarantee of basic rights?	Political unsustainability of ICTs: Contradiction between the rulers and the ruled (dictatorship and war)	To which degree is the social organisation underlying the production or use of ICTs ruled by an elite that excludes others from participation in collective decision-making? To which degree does the use of ICTs foster violence, the violation of basic rights and warfare?
Society: Cultural system	Digital cultural sustainability: Recognition (Questions about online community and eLearning)	To which degree does digital culture enable the development of the human mind, the recognition of identities in society, and the reproduction of the human body?	Digital cultural unsustainability: Contradiction between the cultural elite and everyday people (disrespect and malrecognition)	To which degree does digital culture limit mental development and production, the recognition of identities and the reproduction of the human body?

Conclusion: Towards a Critical, Dialectical Understanding of Sustainability in the Information Society

The dimensions of sustainability do not exist independently, but are interdependent, i.e. a lack of a certain dimension eventually will have negative influences on other dimensions, whereas enrichment of one dimension will provide a positive potential for the enrichment of other dimensions. So for example people who live in poverty are more likely to not show much interest in political participation. Another example is that an unsustainable ecosystem advances an unsustainable society and vice versa: If man pollutes nature and depletes non-renewable natural resources, i.e. if he creates an unhealthy environment, the problems such as poverty, war, totalitarianism, extremism, violence, crime, etc. are more likely to occur. The other way round a society that is shaken by poverty, war, a lack of democracy and plurality, etc. is more likely to pollute and deplete nature. So sustainability should be conceived as being based on dialectics of ecological preservation, human-centred technology, economic equity, political participation, and cultural recognition. These dimensions are held together by the logic of co-operation, i.e. the notion that systems should be designed in ways that allow all involved actors to benefit. Co-operation is the unifying and binding force of a participatory, co-operative, sustainable information society. The logic of co-operation dialectically integrates the various dimensions of sustainability.

The WSIS Civil Society Plenary (2005) argues that in the WSIS process, civil society interests were not adequately taken into account (for a critique of WSIS see also Servaes and Carpentier 2006).

> "Internet access, for everybody and everywhere, especially among disadvantaged populations and in rural areas, must be considered as a global public good. [...] The WSIS documents also mostly focus on market-based solutions and commercial use. Yet the Internet, satellite, cable and broadcast systems all utilize public resources, such as airwaves and orbital paths. These should be managed in the public interest as publicly owned assets through transparent and accountable regulatory frameworks to enable the equitable allocation of resources and infrastructure among a plurality of media including community media".
>
> (WSIS Civil Society Plenary 2005, 4, 12)

In its own declaration – that is very different from the official dualistic WSIS outcome documents – the WSIS Civil Society Plenary (2003) argues for an information society that is based on 34 inclusive principles. Among them are the promotion of free software and the establishment of a public domain of global knowledge that challenges intellectual property. The focus is on public goods and redistribution. The Plenary stresses that distributive justice is needed and that economic resources should not simply be produced within economic growth models, but need to be redistributed:

> "We aspire to build information and communication societies where development is framed by fundamental human rights and oriented to achieving a more equitable distribution of resources, leading to the elimination of poverty in a way that is non-exploitative and environmentally sustainable".
>
> (WSIS Civil Society Plenary 2003, 3)

This chapter has shown that sustainability in the information technology context has played an ideological role that aims to advance a neoliberal policy framework that conceives ICTs as a realms of private capital accumulation and advances the commodification of communications and society. The question that arises in this context is if from a critical theory perspective, the sustainability concept should therefore be discarded or not. The view advanced in this chapter is that a critical social theory should provide an ideology critique of information technology sustainability, but at the same time not discard, but sublate the sustainability concept into a critical notion of (un)sustainable information technology sustainability. Such a concept stands in the context of the quest for an alternative framework for the information technology that goes beyond capital accumulation and aims to advance communications as a commons.

Notes

1 http://www.ispreview.co.uk/index.php/2014/11/eu-unveils-gbp250bn-investment-plan-infrastructure-broadband.html.
2 http://ec.europa.eu/priorities/jobs-growth-and-investment_en.
3 See: Mobile networks hand small fortune to shareholders – but little to taxpayers. *The Guardian Online*, July 31, 2013. Vodafone-Verizone deal: Margaret Hodge raises alarm over tax loss. *The Guardian Online*, September 2, 2013. Tax breaks used by mobile phone networks face scrutiny. *The Guardian Online*, July 31, 2013.
4 See: Luxembourg Tax Files Leaks: Tech Companies, http://www.icij.org/project/luxembourg-leaks/explore-documents-luxembourg-leaks-database (accessed on 15 February 2016).
5 The following industries were for this purpose classified as information industries: advertising, broadcasting & cable, communications equipment, computer & electronics retail, computer hardware, computer services, computer storage devices, consumer electronics, electronics, Internet retail, printing & publishing, semiconductors, software & programming, telecommunications.
6 Six of biggest ten firms pay no UK corporation tax. *The Sunday Times*, January 31, 2016, p. 14.

References

Bijker, Wiebe E., Thomas P. Hughes and Trevor Pinch, eds. 1987. *The Social Construction of Technological Systems*. Cambridge, MA: The MIT Press.

European Commission. 2016. *The Investment Plan for Europe State of Play*. http://ec.europa.eu/priorities/sites/beta-political/files/ip-eu-state-of-play-jan-2016_en.pdf

European Commission. 2015. *Connectivity: Broadband Market Developments in the EU. Digital Agenda Scoreboard 2015*. http://ec.europa.eu/newsroom/dae/document.cfm?action=display&doc_id=9929

European Commission. 2010. *A Digital Agenda for Europe*. http://eur-lex.europa.eu/legal-content/EN/TXT/PDF/?uri=CELEX:52010DC0245&from=EN

Fasenfest, David. 2007. Critical Sociology. In *21st Century Sociology. A Reference Handbook. Volume Two*, ed. Clifton D. Bryant and Dennis L. Peck, 17–23. Thousand Oaks, CA: Sage.

Feenberg, Andrew. 2006. What is Philosophy of Technology? In *Defining Technological Literacy: Towards an Epistemological Framework*, ed. John R. Dakers, 5–16. Dordrecht: Springer.

Forbes. 2000. https://www.forbes.com/sites/liyanchen/2015/05/06/the-worlds-largest-companies/

Fuchs, Christian. 2017. Critical Social Theory and Sustainable Development: The Role of Class, Capitalism and Domination in a Dialectical Analysis of Un/Sustainability. *Sustainable Development* 25 (5): 443–458.

Fuchs, Christian. 2016. *Reading Marx in the Information Age. A Media and Communication Studies Perspective on "Capital Volume I"*. New York: Routledge.

Fuchs, Christian. 2014a. The Dialectic: Not Just the Absolute Recoil, but the World's Living Fire That Extinguishes and Kindles Itself. Reflections on Slavoj Žižek's Version of Dialectical Philosophy in "Absolute Recoil: Towards a New Foundation of Dialectical Materialism". *tripleC: Communication, Capitalism & Critique* 12 (2): 848–875.

Fuchs, Christian. 2014b. *Digital Labour and Karl Marx*. New York: Routledge.

Fuchs, Christian. 2011. *Foundations of Critical Media and Information Studies*. New York: Routledge.

Fuchs, Christian. 2010. Theoretical Foundations of Defining the Participatory, Co-operative, Sustainable Information Society (PCSIS). *Information, Communication, and Society* 13 (1): 23–47.

Furhoff, Lars. 1973. Some Reflections on Newspaper Concentration. *Scandinavian Economic History Review* 21 (1): 1–27.

Habermas, Jürgen.1968/1989. Technology and Science as "Ideology". In *Jürgen Habermas on Society and Politics: A Reader*, ed. Steven Seidman, 237–265. Boston, MA: Beacon Press.

Harvey, David. 2005. *A Brief History of Neoliberalism*. Oxford: Oxford University Press.

Hilty, Lorenz M. and Thomas F. Ruddy. 2010. Sustainable Development and ICT Interpreted in a Natural Science Context. The Resulting Research Question for Social Sciences. *Information, Communication & Society* 13 (1): 7–22.

Hofkirchner, Wolfgang. 2013. *Emergent Information: A Unified Theory of Information Framework.* Singapore: World Scientific.

Horkheimer, Max and Theodor W. Adorno. 2002. *Dialectic of Enlightenment.* Stanford, CA: Stanford University Press.

Illich, Ivan. 1973. *Tools for Conviviality.* Glasgow: Fontana/Collins.

Li-Hua, Richard. 2009. Definitions of Technology. In *A Companion to the Philosophy of Technology,* ed. Jan Kyrre Berg Olsen, Stig Andur Pedersen and Vincent F. Hendricks, 18–22. Malden, MA: Wiley-Blackwell.

Lukács, Georg. 1971. *History and Class Consciousness.* London: Merlin.

MacKenzie, Donald and Judy Wajcman, eds. 1999. *The Social Shaping of Technology.* Maidenhead: Open University Press.

Marcuse, Herbert. 1969. *An Essay on Liberation.* Boston, MA: Beacon Press.

Marcuse, Herbert. 1964. *One-Dimensional Man.* Boston, MA: Beacon Press.

Mulder, Karel, Didac Ferrer and Harro van Lente, eds. 2011. *What is Sustainable Technology? Perceptions, Paradoxes and Possibilities.* Sheffield: Greenleaf Publishing.

Noam, Eli. 2009. *Media Ownership and Concentration in America.* Oxford: Oxford University Press.

Reydon, Thomas A.C. 2012. Philosophy of Technology. In *Internet Encyclopedia of Philosophy,* ed. James Fieser and Bradley Dowden. http://www.iep.utm.edu/technolo/

Saad-Filho, Alfredo and Deborah Johnston, ed. 2005. *Neoliberalism: A Critical Reader.* London: Pluto Press.

Schot, Johan and Arie Rip. 1996. The Past and Future of Constructive Technology Assessment. *Technological Forecasting and Social Change* 54 (2/3): 251–268.

Servaes, Jan and Nico Carpentier, eds. 2006. *Towards a Sustainable Information Society. Deconstructing WSIS.* Bristol: Intellect.

Sohn-Rethel, Alfred. 1978. *Intellectual and Manual Labour: A Critique of Epistemology.* London: Macmillan.

Stiglitz, Joseph E., Amartya Sen and Jean-Paul Fitoussi. 2010. *Report by the Commission on the Measurement of Economic Performance and Social Progress.* Paris: Commission on the Measurement of Economic Performance and Social Progress.

Therborn, Göran. 2012. The Killing Fields of Inequality. *International Journal of Health Services* 42 (4): 579–589.

United Nations General Assembly. 2015. *Outcome Document of the High-Level Meeting of the General Assembly on the Overall Review of the Implementation of the Outcomes of the World Summit on the Information Society.* http://workspace.unpan.org/sites/Internet/Documents/UNPAN95735.pdf

United Nations Human Development Report 2015 (UNHDR 2015). New York: United Nations Development Programme.

Williams, Raymond. 1983. *Keywords. A Vocabulary of Culture and Society.* New York: Oxford University Press.

References

Williams, Raymond. 1976. *Communications*. Harmondsworth: Penguin.

Winner, Langdon. 1986. *The Whale and the Reactor: A Search for Limits in an Age of High Technology*. Chicago, IL: University of Chicago Press.

World Summit on the Information Society. 2005. *Tunis Agenda for the Information Society*. http://www.itu.int/net/wsis/docs2/tunis/off/6rev1.html

World Summit on the Information Society. 2003. *WSIS Geneva Declaration of Principles: Building the Information Society: A Global Challenge in the New Millennium*. http://www.itu.int/net/wsis/docs/geneva/official/dop.html

WSIS+10 High-Level Event. 2014. *WSIS Outcome Documents*. Geneva: ITU.

WSIS Civil Society Plenary. 2005. *Much More Could Have Been Achieved: Civil Society Statement on WSIS*. http://www.worldsummit2005.de/download_en/WSIS-CS-summit-statement-rev1-23-12-2005-en.pdf

WSIS Civil Society Plenary. 2003. *"Shaping Information Society for Human Needs": Civil Society Declaration to the World Summit on the Information Society*. http://www.itu.int/net/wsis/docs/geneva/civil-society-declaration.pdf

Chapter Nine
Theoretical Foundations of Defining the Participatory, Co-operative, Sustainable Information Society (PCSIS)

9.1 Introduction: A Theoretical Model of Society

Sustainable information society (SIS), sustainable knowledge society, sustainable productive information society, sustainable networked knowledge society, planetary sustainable information and knowledge society, participatory information society, inclusive information society, and information society for all: these are some of the buzzwords that have in recent years been employed in the academic and the political discourses on which society is desirable. Overall, these discourses signify a shift toward the view that not just any information society that is brought about by the diffusion of digital networked information and communication technologies (ICTs) is needed, but an information society that is actively shaped by humans in order to gain desirable qualities. Normative judgements have become more important. But these discourses are also fragmented and are lacking a theoretical foundation that tries to give concise definitions of the categories in use. A systematic theoretically grounded comparison of such categories is still missing.

The task of this chapter is to contribute to the elimination of this deficit by constructing a comparative typology of approaches that are grounded in social theory. Furthermore, based on this discussion a dialectical notion that describes an information society that is normatively desirable for the author and that is suggested for broader consideration is introduced. So the second task of this chapter is to theoretically ground a notion of the participatory, co-operative, SIS (PCSIS). The importance of this undertaking is justified by the fact that during the past few years the insight has become common that not just any type of information society is needed, but an information society for all. In this context, the notions of participation, co-operation, and sustainability have become important in information society discourse.

DOI: 10.4324/9781003279488-11

First, the theoretical background is outlined (Section 9.1), then a typology of approaches on PCSIS will be introduced (Section 9.2), and finally some conclusions are drawn (Section 9.3). Methodologically, this chapter is based on social theory construction and dialectical thinking. By identifying two poles that are important for social theory construction (base and superstructure), a criterion that allows the distinction of various theoretical approaches on the notion of the SIS is introduced. Four possible relations between these two poles are considered as grounding four different approaches. The last of these four ones is considered as a dialectical approach in the sense that the two poles are different (each is characterised by the absence of qualities of the other, cf. Bhaskar 1993), but at the same time related, connected, interdependent, and mediated (Hegel 1830, §116). Two moments are connected, which are at the same time different and connected and are encroaching on each other (Holz 2005). The approach introduced by the author is a dialectical one and can be seen as a synthesis of the other three introduced approaches.

Models of society that privilege one part over other parts, such as economism, politicism, or culturalism, are not able to explain phenomena that show a relative autonomy. So for example models of society that reduce explanations to the economy, cannot explain why protest movements can emerge both in situations of relative economic stability or instability (compare for instance the economic conditions in the era of the rise of the Nazi movement with those of the era of the 1968 students' movement). Models of society that see society as being composed of independent subsystems, such as Luhmann's (1984) theory of functional differentiation, face the problem of explaining phenomena that are characteristic for the global network society. So they for example cannot grasp that today economic logic influences and dominates large parts of society. In contrast to reductionistic and relativistic social theories, dialectical social theories have proved successful in conceiving society as being composed of relative autonomous subsystems that all have their own specificity, but nonetheless depend on each other and influence each other. The subsystems are conceived as distinct and at the same time mutually interdependent, which is the fundamental logical figure of dialectical thinking.

Society can be conceived as consisting of interconnected subsystems that are not independent and based on one specific function they fulfill, but are open, communicatively interconnected, and networked. As subsystems of a model of society one can conceive the ecological system, the technological system, the economic system, the political system, and the cultural system (Fuchs 2008c, cf. Figure 9.1). Why exactly these systems? In order to survive, humans in society have to appropriate and change nature (ecology) with the help of technologies so that they can produce resources that they distribute and consume (economy), which enables them to make collective decisions (polity), form values,

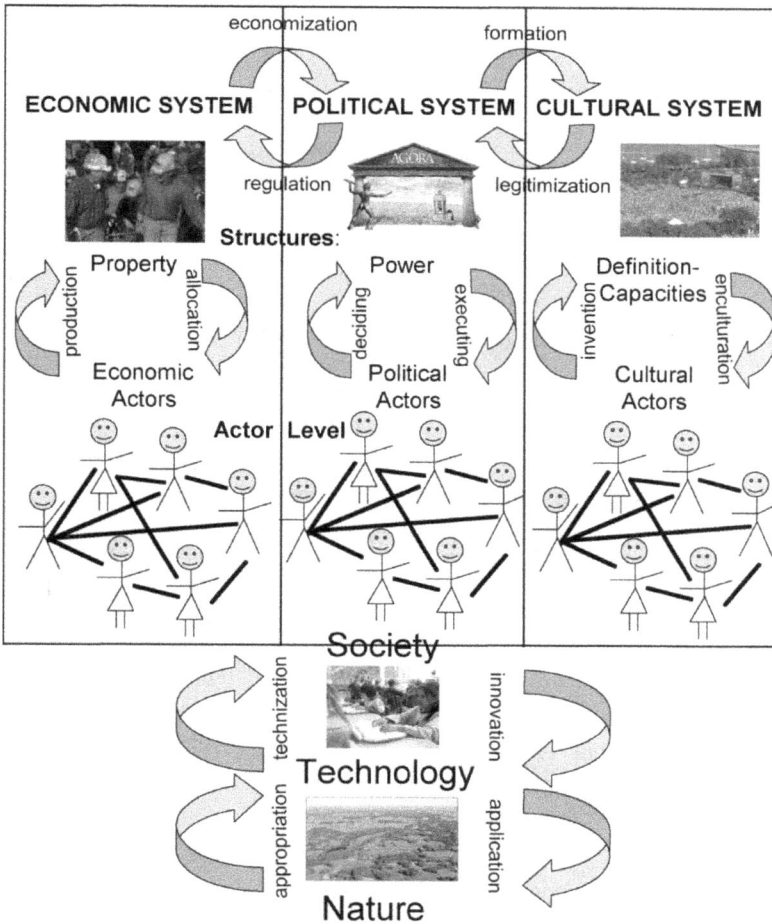

FIGURE 9.1 Society as a dynamic, dialectical system.

and acquire skills (culture). The core of this model consists of three systems (economy, polity, and culture). This distinction can also be found in other contemporary sociological theories: Giddens (1984, 28–34) distinguishes between economic institutions, political institutions, and symbolic orders/modes of discourse as the three types of institutions in society. Bourdieu (1986) speaks of economic, political, and cultural capital as the three types of structures in society. Jürgen Habermas (1981) differs between the lifeworld, the economic system, and the political system.

Each of these three systems is shaped by human actors and social structures that are produced by the actors and condition the actors' practices. Each subsystem is defined and permanently re-created by a reflexive loop that productively interconnects human

TABLE 9.1 An overview of structures in society

Type of structure	Structure	Definition
Ecological structures	(Natural) resources	Physical matter that is extracted in labour processes from nature and that is changed by human activities.
Technological structures	Tools	Artefacts, means, methods, and skills of action that are used by humans in order to try to achieve defined goals.
Economic structures	Property	Goods and resources that are produced, distributed, and used by humans for satisfying defined needs.
Political structures	Power	The capacity and means for influencing collective decisions according to one's own will.
Cultural structures	Definition-capacities	The capacity to define and acquire values, skills, and practices that shall give meaning to life and help re-create human minds and bodies.

actors and their practices with social structures. An overview of the qualities of structuring and structured structures in society is given in Table 9.1.

The economic system can only produce goods that satisfy human needs by human labour power that makes use of productive and communication technologies in order to establish social relations and change the state of natural resources. The latter are transformed into economic goods by the application of technologies to nature and society in labour processes. The economy is based on the dialectic of natural resources and labour that is mediated by technology. We can therefore argue that socially transformed nature and technology are aspects of the economic system.

This allows us to make a distinction between the base and the superstructure of society. The economic base is constituted by the interplay of labour, technology, and nature so that economic goods are produced that satisfy human needs. The superstructure is made up by the interconnection of the political and the cultural system, so that immaterial goods emerge that allow the definition of collective decisions and societal value structures. Does it make sense to speak of base (nature, technology, and economy) and superstructure (polity, culture) in society, or does this mean that one reduces all social existence to economic facts? The superstructure is not a mechanic reflection, that is, a linear mapping, of the base, that is, the relations and forces of production. It cannot be deduced from or reduced to it. All human activity is based on producing a natural and social environment; it is in this sense that the notion of the base is of fundamental importance. We have to eat and survive before we can and in order to enjoy leisure, entertainment, arts, and so on. The base is a precondition, a necessary, but not a sufficient condition for the superstructure. The superstructure is a complex, nonlinear creative reflection of the base, the base a complex, nonlinear creative reflection of the superstructure. This means that both levels are recursively linked and produce each other. Economic practices and structures trigger political and cultural processes. Cultural and political practices and structures trigger economic processes. The notion of creative reflection grasps the dialectic of chance and necessity/indetermination and determination that shapes the relationship of base and superstructure. There is not a content of the superstructure that is "predicted, prefigured and controlled" by the base; the base as Raymond Williams in his famous paper on *Base and Superstructure* has argued "sets limits and exerts pressure" on the superstructure (Williams 2001, 165). Stuart Hall (1983) has in this context spoken of a determination in the first instance exerted by the economic system on superstructures.

9.2 Concepts of a Sustainable Information Society (SIS)

Wolfgang Hofkirchner (2002) has introduced a typology of four worldviews that are based on the potential relationships between two categories: reductionism establishes identity by eliminating the difference for the benefit of the smaller, less differentiated part; projectionism establishes identity by eliminating the difference for the benefit of the larger, more differentiated side; dualism eliminates identity by establishing a difference of the two sides, it is a disjunctive approach; finally, dialectical thinking integrates the two sides so that the two sides have different and identical aspects; they yield a unity in diversity. The advantage of using this typology instead of other approaches is that it covers all logical possibilities of how two entities can be related.

Applying Hofkirchner's typology to the relationship of base and superstructure allows us to classify definitions of a participatory/co-operative/SIS. The base is less differentiated than the superstructure because all superstructural phenomena have economic aspects, whereas not all economic phenomena have political and cultural aspects. Hence the superstructure is more differentiated and builds upon the base. There are reductionistic, projective, dualistic, and dialectical approaches. Reductionistic approaches reduce sustainability to the economic base, i.e. they see economic, technological, or ecological aspects as the determining factors, the superstructure is deduced from the base. Projective approaches consider political or cultural aspects as the sole determining factors of sustainability, they give priority to the superstructure, and the base is derived from the superstructure. Dualistic approaches assert the existence of a variety of dimensions of sustainability, but they consider these dimensions as being independent. Dialectical thinking conceives sustainability as on the one hand multidimensional and on the other

hand interdependent. The various dimensions are seen as having their own specific relative autonomies, but as being at the same time causally related in complex ways, mutually constituting and influencing.

The reason why it might be problematic to speak of a "sustainable information society"
or a "participatory information society" is that there is evidence that huge gaps in wealth
are characteristic of contemporary society and indicate the existence of a capitalist class
system that is euphemised or blanked out by the concepts of sustainability and participation (for a detailed macroeconomic statistical analysis compare Fuchs 2008c). Many
Western countries have relatively high poverty rates that are well above 10% of the
population. So for example in the USA, relative poverty rose from 21.0% in 1974 to
24.1% in 2004 (Data: Luxembourg Income Study). In the UK, it increased from 12.4% in
1969 to 19.2% in 2004 (ibid.). Also, income inequality measured by the Gini coefficient
has increased in many countries during the past decades. So for example in the USA, the
Gini coefficient rose from 30.1 in 1979 to 37.2 in 2004 and in the UK in the same years
from 27.0 to 34.5 (ibid.). Branko Milanovic (2002, 2007a, 2007b) has calculated global
inequality based on household income surveys from 91 countries. He has calculated an
increase in Gini inequality for the period of 1988–1993 from 62.5 to 66 and from 66 to 70
for the period of 1993–2003. The developed world accounts for approximately 25% of
the world's population (United Nations Human Development Report (UNHDR) 2008), but
has since the early 1970s accounted for the majority of the world's wealth (Figure 9.2).
The least developed countries' share has dropped from above 3% to a little above 1%
in the time from 1980 to 2007 (UNHDR 2008). The share of Sub-Saharan countries has
remained continuously below 1% during the same time (ibid.). One big change in the

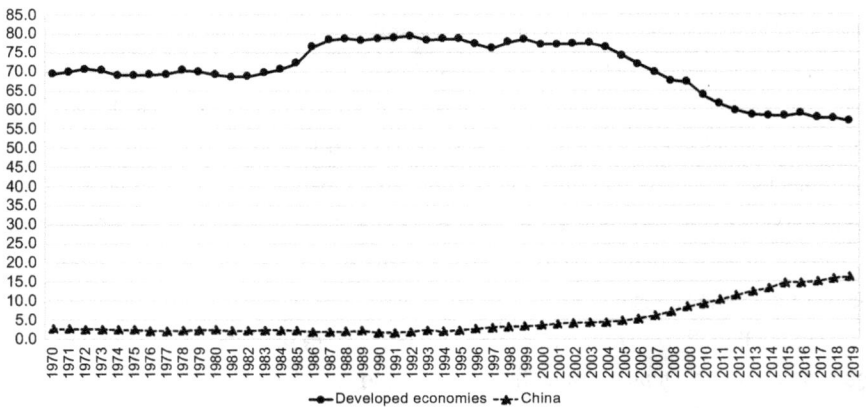

FIGURE 9.2 Share of developed economies in world GDP.

world economy has been the strong growth of the Chinese economy. The UNCTAD data, in Figure 9.2, do not include China under the developed economies. But its share of the world's gross domestic product has risen from 2.7% in 1970 to 16.2% in 2019, which is the major factor that has resulted in a drop of the developed economies share from 69.4% in 1970 to 56.8% in 2019. If we include China among the developed economies, then the share of developed economies in the global GDP has risen from 72.1% in 1970 to 73.0% in 2019. Poor parts of the world such as the Sub-Saharan countries have remained extremely poor during this time period.

In Western countries, productivity increases have resulted in the past decades in continuously rising profit shares, but at the same time in a drop of the wage rate, which shows that capital accumulation has been driven by relative drops in total wages. In the EU15 countries, productivity increased from an index value of 49.7 in 1960 to one of 104.6 in 2009 (Annual Macro-Economic Database). During the same time, total annual corporate profits increased from €100.0 billion to €2979.8 billion and the wage share dropped from 62.7 to 57.3 (ibid.). In the USA, productivity increased from an index value of 60.6 in 1960 to one of 105.7 in 2005 (ibid.). During the same time, total annual corporate profits increased from US$131 billion in 1960 to US$3,594.8 billion in 2009, and the wage share dropped from 65.3 to 60.8 (ibid.). In Japan, productivity increased from an index value of 36.4 in 1960 to one of 112.8 in 2009 (ibid.). During the same time total annual corporate profits increased from ¥6.6 billion to ¥97.2 billion and the wage share dropped from 73.2 to 58.5 (ibid.). Is it really likely and reasonable to assume that societies characterised by such class divisions have potentials for becoming "sustainable" in the near future?

Table 9.2 gives on overview of the approaches and examples that are discussed in Sections 9.2.1–9.2.4. In the discussion in the succeeding chapters, examples for each of the four kinds of approaches are given.

9.2.1 Reductionistic SIS Definitions: Base without Superstructure

Reductionistic approaches see ecological or technological or economic changes as the sole driving forces of a SIS.

O'Donnell, McQuillan and Malina (2003, 26ff; cf. also O'Donnell 2001) define an inclusive information society as a society that ensures that all citizens (especially the elderly, women at home, the disabled, farmers, the unemployed, etc.) have the opportunity to

TABLE 9.2 Approaches on the sustainable information society

Type of approach on sustainable information society	Examples	Description of approach
Reductionism	Britton (1996), Commission of the European Communities (2005), Hilty (2000), Lisbon European Council (2000), O'Donnel, McQuillan, Malina (2003), O'Donnell (2001),	Ecology, economy, or technology is considered as the driving forces of a sustainable information society.
Projectionism	Macintosh (2004, 2006)	Polity and/or culture are seen as the determinant forces of a sustainable information society.
Dualism	Carrelli et al. (2000), Club of Rome (2003), Club of Rome and the Factor 10 Institute (2002), Commission of the European Communities (2002, 2006), European Commission (1998), eEurope (2002a, 2002b), eEurope (2005), Information Society Forum (1998, 2000), Radermacher (2004), Schauer (2003), World Summit on the Information Society (2005)	Multiple dimensions and goals of a sustainable information society are identified, but not causally related to each other.
Dialectic	Fuchs (2008a, 2008b, 2008c), Göhring (1999), Heinrich Böll Foundation (2003a, 2003b), López-Ospina (2003), World Summit on the Information Society Civil Society Plenary (2005)	Multiple interrelated dimensions and goals of a sustainable information society are identified, existing contradictions of these dimensions are analysed, and necessary changes are conceived as integral, interdependent, and systemic.

use ICTs to improve the quality of their lives and communities (community-building and -maintenance, eCommerce, eBusiness, eLearning, eLeisure, eHealth, and eGovernment), to contribute to a knowledge-based economy and society (improving human capital and technology-related skills, foster ICT-related economic growth, increase the use of ICTs, special support for ICT learning and skills for disadvantaged individuals and rural areas), and to engage with Government services and participate in the democratic process and that civil society is engaged with the help of ICTs in ICT training, employment, democratic participation, online content production, and the building of social capital and trust in ICTs.

This definition involves economic, political, and cultural aspects, its main problem is that its four aspects are strongly overlapping and hence have no analytical discriminatory power. It is a very technology-centred definition, led by the belief that technology access and skills alone suffice to improve the lives of all. What is missing is the insight that technology support needs to be combined with social transformations toward participatory social systems. The concept of eInclusion provided by O'Donnell et al. is an example of a technodeterministic-reductionistic definition of SIS.

Lorenz Hilty defines a SIS in purely ecological terms, i.e. with a focus on environmental protection. He argues that the sustainability aspect of ICTs is "how they could help reduce the material intensity of economic processes" (Hilty 2000, 6). His approach is an ecological reductionistic one.

Another ecological reductionistic view is presented in a special issue of the *Journal of World Transport Policy & Practice* (Britton 1996). It is argued that ICTs support sustainability because they are revolutionary in the respect that they "(a) involve relatively small amounts of material resources, and (b) permit substantial dematerialisation in many domains" (Britton 1996, 12). Ecological benefits could be achieved by implementing sustainable transport as a bridging strategy that would have positive effects upon society as a whole. "Even without some form of collective guidance, the Information Society is likely to have significant de-materialization impacts, which will work in the direct of more sustainable behavior" (Britton 1996, 12). Nonetheless, conscious political action would be necessary in order to tackle unsustainable tendencies. This approach is reductionistic in the sense that it is focused on ecological and transport issues and neglects a variety of other issues that are seen as being derived from transport issues.

The development initiative i2010 of the European Commission is oriented on purely economic issues, arguing that what is needed today is an "information society for growth and employment" (Commission of the European Communities 2005). The main objectives of i2010 are technological progress ("a single European information space"), the advancement of research and innovation in ICTs, and inclusiveness (ibid.). The economic goal formulated in the Lisbon strategy "to become the most competitive and dynamic knowledge-based economy in the world capable of sustainable economic growth with more and better jobs and greater social cohesion" (Lisbon European Council 2000) by 2010 has been the driving force of all such EU initiatives. In the eEurope strategy, economic goals were seen as being achievable by investment in the economy, polity, culture, and welfare. This was a dualistic strategy that identified multiple separate goals. The difference that has emerged with i2010 is that now a rather strictly economic strategy has been introduced, defining economic goals (growth, employment) as the most important ones. The EU has shifted from a dualistic toward an economic reductionistic strategy.

9.2.2 Projectionistic SIS Definitions: Superstructure without Base

Projectionistic approaches see superstructures (polity and/or culture) as the determining forces of sustainability. They are the least frequently found approaches in the literature.

One example is the notion of participation underlying the concept of eParticipation as defined by Ann MacIntosh. The approach focuses on ICTs for advancing the inclusion of citizens in political decision-making. eParticipation is defined as "the use of information and communication technologies to broaden and deepen political participation by enabling citizens to connect with one another and with their elected representatives" (MacIntosh 2006, cf. also the contributions in DEMO-net 2006). For MacIntosh (2004), eParticipation consists of the three aspects of E-enabling, E-engaging and E-empowering citizens to participate in politics. The focus is strictly on political processes, participation in cultural or economic, or technological processes is neglected. This approach is a projectionistic politicism.

9.2.3 Dualistic SIS Definitions: Separating Base and Superstructure

Dualistic approaches define multiple goals and dimensions of SIS, but do not consider if these goals are compatible and if and how they are causally linked. Dualistic models are the ones that can be found most frequently in the literature.

The European Commission has advanced a dualistic view of the SIS by arguing that ICTs support economic growth, social progress, and environmental sustainability:

> "Investing in knowledge is certainly the best, and maybe the only, way for the EU to foster economic growth and create more and better jobs, while at the same time ensuring social progress and environmental sustainability. In other words, it is Europe's chance to strengthen its model of society".
>
> (Commission of the European Communities 2006, 2)

It has stressed the importance of economic growth without considering that economic accumulation has been bringing about income inequalities during the past decades.

> "It is clear that with the Information Society, new opportunities are emerging which will help to achieve both global environmental sustainability and continued economic growth; to achieve social goals of employment growth and local

community development <u>within</u> a free market framework; and to enable greater access to work, services and mobility <u>without</u> congestion".

(European Commission 1998, 9)

The aim of the eEurope initiative was an information society for all, understood as bringing "everyone in Europe – every citizen, every school, every company – online as quickly as possible" (Commission of the European Communities 2000a, 2000b). In the first phase of this initiative, the focus was on advancing e-commerce, access for youth, researchers, students, the disabled, smart cards, eHealth, eTransport, and eGovernment). In the second phase, the focus was on advancing eGovernment, eLearning, eHealth, eBusiness, information infrastructure, and security (Commission of the European Communities 2002). eEurope was a dualistic strategy that identified multiple separate goals without taking into consideration the issue of compatibility of the various goals.

A similar dualistic view is advanced by the World Summit on the Information Society (2005, 10), which argues that ICTs can sustain "economic growth, job creation and employability and improving the quality of life of all".

The goal of the Global Marshall Initiative, headed by Franz Josef Radermacher, is the implementation of a worldwide eco-social market economy that is simultaneously oriented on the growth of economic value-adding and on worldwide social, cultural, and ecological solidarity (Radermacher 2004, 47, 48). The worldview underlying this conception is termed ordoliberalism and is based on five aspects: international contracts, environmental protection, social balance, cultural diversity and tolerance, and "the intention to further open markets internationally and co-finance development matters on the content of a common observance of standards" (62). Needed would also be further market openings by developed countries for the benefit of the rest of the world (59). "As has been illustrated, an Eco-Social Market Economy relies on the power of markets and competition, but it has to be subject to an eco-social regulatory framework at the same time" (66). Radermacher wants to extend and intensify the dominance of economic logic (further opening markets) and at the same time advance non-economic social and ecological benefits. He does not recognise that the instrumental reason underlying economic logic has produced many of the problems that humanity is facing today (as will be shown later in this paper). An eco-social world market economy from this perspective is a contradictio in adiecto and what is needed is not an "Eco-Social World Market Economy which links markets and competition to high standards ensuring the welfare of all human beings" (166), but participation and co-operation as sustainable alternatives to market logic and competition.

Concepts of a Sustainable Information Society (SIS)

A publication by the Information Society Forum, in which Radermacher is very influential, suggests that important economic measures for achieving a SIS are the "extension of liberalization and competition policy to local access networks" and increasing the "availability of risk capital for entrepreneurs" (Information Society Forum 2000). Economic crisis tendencies such as the South-East Asian crisis in 1997, the new economy crisis in 2000, and the housing market crisis in the USA in 2008 that triggered a new world economic crisis, have shown that speculative finance capital based on liberalised commodity and financial markets increases economic vulnerability and is a risk for general wealth. Given the crisis-ridden nature of finance capitalism, it is almost ridiculous to make these recommendations.

Radermacher's view, which is also the one of the Information Society Forum, is that ICTs have a potential to bring about dematerialisation, and that the latter simultaneously helps to achieve sustainability and economic growth (Information Society Forum 1998, 93, 95). Dematerialisation has thus far not been successful and it could indeed be detrimental to economic growth if reusable ICT equipment were introduced (Fuchs 2008b). In another publication, Radermacher's dualistic view is legitimated by publishing together with important persons such as information society researcher Jan Van Dijk (Carrelli et al. 2000). One mechanism that is specifically stressed in this chapter is the one of global trading in pollution rights. The market mechanisms that have caused unsustainable development (as will be argued with the help of statistics later in this chapter) are considered as solutions to the created problems – a contradictio in adiecto. The general argument is that "free markets" must be "complemented" (50) by social, cultural, political, and ecological framework. It is not taken into account that free markets might hinder such frameworks and hence need not be complemented, but driven back and contested (as will be argued later in this chapter). Another problematic aspect of this and other publications by Radermacher is that global population growth is considered as a source of unsustainable development and the shrinking of the world population as a goal. It is not taken into account that population growth is a reaction to global income inequality and that economic productivity has today reached levels that allow a good life for all people worldwide, given there is a primacy of global and national economic redistribution (which is not the case for Radermacher).

Schauer (2003) provides another dualistic approach by arguing that ICTs can advance ecological sustainability by reducing resource consumption, social sustainability by giving equal access to information, cultural sustainability by supporting cultural understanding, and economical sustainability by fostering growth: "Information technology will be the key driver of an economic growth which is decoupled from resource consumption" (Schauer 2003, 32). The question whether economic growth in late-modern society is compatible with social sustainability is not considered. This definition does not see that capitalist development has hindered social equality especially in the last decades (as will be shown below), it treats economic profitability as one major goal besides ecological, social, and cultural issues.

The Club of Rome and the Factor 10 Institute (2002) defines a sustainable networked knowledge society as a society in which ICTs foster entrepreneurship and access to world markets even in the poorest regions of the world and provide higher eco-efficiency of economic growth (social sustainability), ICTs enable global communication that allows the emergence of cultural diversity, respect for human rights, and a global culture of co-operation (cultural sustainability), ICTs support resource-use efficiency, the reduction of toxic anthropogenic material cycles, and the emergence of environmentally sustainable lifestyles (ecological sustainability), and ICTs advance economic growth and profitability (economic sustainability). This approach is dualistic, it argues that "economic sustainability" is a very important dimension and does not see that the current model of the economy and economic growth threatens sustainability. In another publication, the Club of Rome affirms this position by arguing that among various measures also the "liberalization of information and communication network infrastructure and service provision" (Club of Rome 2003, 10) is important.

The reason why we question dualistic approaches is that there is evidence that late-modern society is characterised by a culminating antagonism between economic growth and social and ecological cohesion, economic freedom (of markets) and social equity. Income inequality measured as the relation of the mean income of the upper and the lower quintile has decreased in the years 1995–2000 in the EU15 countries, but it has increased from 4.5 in 2000 to 4.8 in 2005 (Eurostat Online). The higher this measure, the higher the income disparity between the poorest and the richest. In the EU25 countries, it has increased from 4.5 in 2000 to 4.9 in 2005. In 2000, the richest 5% of Europeans owned 35.7% of the worldwide wealth (Davies et al. 2006, Table 10a). The at-risk-of-poverty rate after social transfers measured by 60% of median equivalenced income after social transfers has risen from 15% in 1998 to 16% in 2005 in the EU15 as well as the EU25 countries (Eurostat Online). Income inequality, as measured by the Gini coefficient, has increased from 29 in 1998 to 31 in 2005 in the EU25 countries and from 29 in 1998 to 30 in 2005 in the EU15 countries (Eurostat Online). The in-work at risk of poverty rate for part-time workers was 11% in the EU25 and 10% in the EU15 countries in 2005 (Eurostat Online). The increase in income inequality, job insecurity,

Concepts of a Sustainable Information Society (SIS)

and poverty risk has been accompanied by a polarisation between capital and labour: whereas the average profit rate has increased by 39.4% in the years 1987–2007 in the EU15 countries (net returns on net capital stock, European Commission Annual Macro-Economic Database), the wage share has in the same time span decreased by 7.5% (Compensation per employee as percentage of GDP at current market prices, European Commission Annual Macro-Economic Database). It is hence reasonable to assume that during the past couple of decades economic growth has been accompanied by a rise of relative wage decreases, income inequalities, and poverty risks. Hence we assume that such a form of economic growth, i.e. the unhindered expansion of capital accumulation, is not compatible with social sustainability. The conclusion of many contemporary social analysts is that the dominance of economic logic needs to be driven back in order to achieve sustainability (e.g. Archer 2007; Harvey 2005; Stiglitz 2003) and that systemic alternatives are needed. It can therefore be hypothesised that "economic sustainability" in the sense of the continued expansion of capitalist accumulation is not compatible with social sustainability and that a paradigm shift is needed. Persistent economic growth has been achieved by compromising social sustainability (e.g. by reducing the total wage labour costs and advancing precarious jobs in order to raise profits) and by externalising economic costs to nature. It has been based on the principle of accumulation by dispossession (Harvey 2005). Less profitability and more corporate taxation are needed in order to provide financial means that can be invested in social and ecological sustainability. Economic sustainability hence should not be understood as meaning continuously rising profit rates, but should better be conceived as self-managed ownership, distributive justice, and the advancement of public goods (based on the insight that the commons are produced co-operatively and hence should be owned collectively).

Interestingly, although this alternative view is not dominant (the dualistic approach is the predominant one), it is shared by a number of institutions and authors who have given definitions of SIS. One such organisation is the Heinrich Böll Foundation:

> "Sustainability of knowledge and information means firstly containing the currently dominating trend towards commodification, which is aimed at short-range use and at creating an artificial scarcity of knowledge, although, as a good, it is essentially free; the agents of commodification are not primarily interested in the long-range securing of individual and social development or for freedom in the use of knowledge and information".
>
> (Heinrich Böll Foundation 2003b, 1)

Another one is the United Nations Educational, Scientific and Cultural Organization (UNESCO):

> "Struggling for development is not to ensure that a few get rich at the expense of the rest, or maintaining non-viable companies or institutions. [...] Globalization currently imposed the notion of the market on everything: education, health, communication services, cultural affairs, etc., and political powers can do nothing about this".
>
> (López-Ospina 2003, 38, 129)

Such views stress a balancing of dimensions, which would require decreasing the predominant economic influence on society. They are dialectical instead of dualistic, projectionistic, or reductionistic.

9.2.4 Towards a Dialectical Definition of SIS: Integrating Base and Superstructure

In 1987, the World Commission on Environment and Development (WCED) published the "Brundtland Report" (named after its Chair, the former Prime Minister of Norway, Gro Harlem Brundtland; WCED 1987) that gave much attention to the challenge of overcoming poverty and meeting basic needs and to integrate the environment into economic decision-making. The WCED defined sustainable development as "development that meets the needs of the present without compromising the ability of future generations to meet their own needs" (WCED 1987, 43). Applying this idea to systems design means that with the help of technology individuals, communication processes, organisations, and societies should be managed and designed in ways that allow all three levels to develop in harmony and achieve their own goals without compromising the goals of the other levels or of other actors in the present and the future.

In the discourse on sustainability, there has been a shift from a focus on ecological issues towards the inclusion of broader societal issues. The "triangle of sustainability" introduced by the World Bank has been very important in shifting the discussion on sustainability from purely ecological aspects toward more integrative concepts. Ismail Serageldin, then vice-president of the World Bank, identified an economic, a social, and an ecological dimension of sustainability. "It is not surprising that these concerns reflect the three sides of what I have called the 'triangle of sustainability'-its economic, social, and ecological dimensions" (Serageldin 1995, 17). It has now become very common to

Concepts of a Sustainable Information Society (SIS)

identify an ecological, an economic, a social, and an institutional dimension of sustainability (as e.g. the EU and the UN do). A shift in the meaning of the sustainability notion occurred between the time of the 1992 UN Conference on Environment and Development ("Earth Summit") in Rio de Janeiro, Brazil, and the 2002 World Summit on Sustainable Development (WSSD) in Johannesburg, South Africa. "At the time of Rio, sustainable development was mainly about protecting nature, but now, in the wake of Johannesburg, it is first and foremost about protecting people" (World Summit on Sustainable Development 2002, 22).

If we conceive sustainability as a complex phenomenon, then it includes various aspects that need to be achieved in sustainable social systems, such as individual well-being, security, freedom, and self-determination just like collective dimensions such as wealth for all, social security for all, political participation for all, or health and education for all.

The correspondence of individual, organisational, and societal goals could also be interpreted as a contemporary form of Kant's Categorical Imperative:

> "Act only according to that maxim by which you can at the same time will that it should become a universal law. [...] Act as though the maxim of your action were by your will to become a universal law of nature. [...] Act so that you treat humanity, whether in your own person or in that of another, always as an end and never as a means only".
>
> (Kant 1998, 422, 429)

Treating others with the same logic that one wants have applied to oneself means that there can be no morally privileged logic at any level. But Kant's Golden Rule fails in situations where people are willing to suffer, tolerate violence against them, or to die if they were in the positions of others. Hence one assumption that might need to be added is that the logics employed at the individual, organisational, and the societal level should be guided by the spirit of co-operation and participation. This implies that the logic of co-operation is superior to the logic of competition.

How can the superiority of co-operation to competition be justified? Competition means that certain individuals and groups benefit at the expense of others, i.e. there is an unequal access to structures of social systems. The asymmetric distribution of resources, domination, and exploitation are the typical outcomes of competition. Competition is the dominant organisational structure of modern society, modern society hence is an excluding society. Cooperation is a specific type of communication where actors achieve a shared understanding of social phenomena, make concerted use of resources so that new systemic qualities emerge, engage in mutual learning, all actors benefit, and feel at home and comfortable in the social system that they jointly construct (Fuchs 2008b). Co-operation includes people in social systems, it lets them participate in decisions and establishes a more just distribution of and access to resources. Hence co-operation is a way of achieving and realising basic human needs, competition is a way of achieving and realising basic human needs only for certain groups and excluding others. We argue that co-operation forms the Essence of human society, and that competition estranges humans from their Essence. One can imagine a society that functions without competition, a society without competition is still a society. One cannot imagine a society that functions without a certain degree of co-operation and social activity. A society without co-operation is not a society, it is a state of permanent warfare, egoism, and mutual destruction that sooner or later destroys all human existence. If co-operation is the Essence of society, then a truly human society is a co-operative society, and from this insight emerges the categoric imperative to overthrow all ideas and practices in which man is not considered as the participating centre of society, but treated as enslaved to instrumental structures.

Participation means that humans are enabled by technologies, resources, organisations, and skills to design and manage their social systems all by themselves and to develop collective visions of a better future so that the design of social systems can make use of their collective intelligence (Fuchs 2008b). A participatory social system is a system in which power is distributed in a rather symmetrical way, that is, humans are enabled to control and acquire resources such as property, technologies, social relationships, knowledge, and skills that help them in entering communication and cooperation processes in which decisions on questions that are of collective concern are taken. Providing people with resources and capacities that enable responsible and critical activity in decision-making processes is a process of empowerment; participation is a process of empowering humans.

How are participation, co-operation, and sustainability connected? Participation is structure-oriented, it is a process in which social structures are designed in such a way that individuals are included in the constitution of the social systems they live in and actually take part in these constitution processes. Co-operation is an intersubjective process within participatory structures, participation is a logical and necessary, but not sufficient precondition for co-operation. Co-operation is the social process by which sustainable systems can be produced. Sustainability concerns the long-term form and effects of a social system. Participation means the structural enablement of humans to take part in collective processes, co-operation the intersubjective social process of working together,

Concepts of a Sustainable Information Society (SIS)

TABLE 9.3 Dimensions of sustainability

Dimension	Definition
Ecology: Preservation	Under the condition of ecological preservation, nature is treated by humans in ways that allow flourishing of natural systems, i.e. the autopoiesis of living systems is maintained and not artificially interrupted or destroyed and natural resources are preserved and not depleted.
Technology: Human-Centredness	That technology is human-centred means that technological systems should help humans in solving problems, fit their capabilities, practices, and self-defined needs, support human activities and co-operation, and involve users in definition, development, and application processes.
Economy: Equity	Economic equity means that there is wealth for all, i.e. defined material living standards should be guaranteed for all as a right, nobody should live in poverty, and the overall wealth should be distributed in a fair way so to avoid large wealth and income gaps between the most and the least wealthy.
Polity: Freedom	Freedom can in line with the critical-realist thinking of Roy Bhaskar (1993) be conceived as the absenting of domination, i.e. the asymmetrical distribution of power, so that humans are included and involved in defining, setting, and controlling the conditions of their lives. It is the absenting of constraints on the maximum development and realisation of human faculties. Freedom then means the maximum use and development of what C.B. MacPherson (1973) has termed human developmental power.
Culture: Wisdom	A culture is wise if it allows the universal sharing and co-operative constitution of knowledge, ideas, values, norms, and sets standards that allow literacy and the attainment of educational skills for all, physical and mental health of all, the maximisation of lifetime in health for all, communicative dialogue in which all voices are heard and influential, a culture of understanding that allows finding common values without compromising difference (unity in diversity), the experience of entertainment, beauty, the diversity of places, mental challenge and diversity, physical exercise for all, and building communities, relations, love, and friendships for all.

and sustainability the long-term condition and effects of social systems so that all benefit and can lead a good life. Abstractly spoken, a PCSIS is a society that guarantees a good life for all. A PCSIS is a society in which knowledge and technology are together with social systems shaped in such ways that humans are included in and self-determine their social systems collectively, interact in mutually benefiting ways, and so bring about a long-term stability that benefits all present and future generations and social groups. Table 9.3 shows the various dimensions of such a society.

The dimensions of sustainability do not exist independently, but are interdependent, i.e. a lack of a certain dimension eventually will have negative influences on other dimensions, whereas enrichment of one dimension will provide a positive potential for the enrichment of other dimensions. So for example people who live in poverty are likely to not show much interest in political participation. Another example is that an unsustainable ecosystem advances an unsustainable society and vice versa: if man pollutes nature and depletes non-renewable natural resources problems, i.e. if he creates an unhealthy environment, problems such as poverty, war, totalitarianism, extremism, violence, crime, etc. are more likely to occur. The other way round a society that is shaken by poverty, war, a lack of democracy and plurality, etc. is more likely to pollute and deplete nature. So sustainability should be conceived as being based on dialectics of ecological preservation, human-centred technology, economic equity, political freedom, and cultural wisdom. These dimensions are held together by the logic of co-operation, i.e. the notion that systems should be designed in ways that allow all involved actors to benefit, co-operation is the unifying and binding force of a PCSIS, it dialectically integrates the various dimensions.

Elements of dialectical approaches on SIS have thus far been marginalised by the dominance of dualistic views. Nonetheless, there are some exceptions. So e.g. the UNESCO is calling for a planetary sustainable information and knowledge society (López-Ospina 2003). It argues for turning away from the pure focus on economic logic and towards a balanced view that takes into account integrative human rights. The goal is a society that realises for all the right to life, right to political participation, right to legal protection, right to freedom, right to benefit of progress to all (economic, social, and cultural participation), right to minimal income for all human beings, right to subsistence income and employment, right to education, right to health, right to sexual and reproductive rights, right to nutrition and food security, right to a healthful environment, right to housing and equitable human settlements (López-Ospina 2003, 180). This view is integrating the ecological, economic, political, cultural, and social dimensions of human existence. There is a stress that economic interests are currently privileged and should be driven back in order to advance advantages for all. So the causal relation between the various dimensions is taken into account (other than in SIS dualism), which results in a call for the decolonisation of society by economic logic. This would also apply for ICTs that according to UNESCO should not be used for purely economic ends, but for fostering planetary sustainability that benefits all humans.

> "Considering new information and communication technologies, governed at present and since their outset by the rules of the market and stock markets, must be lightened in terms of management and international democratic governance by firmly incorporating ethical principles and values that will recognize that it is only by seeking intellectual, spiritual and cultural progress for all peoples that humankind can be prepared for the advent of a more balanced,

Concepts of a Sustainable Information Society (SIS)

equitable, fair world, to assure a good life for all [...] Consequently, information and communication technologies must be used and managed in a society
in order to humanize and democratize thought in society, rather than to en-
hance economic profitability and efficiency, achieved for better or for worse
using sophisticated administrative and management programs grounded in
different realities from those that were the basis for their original creation".

(70–71)

There is a stress on the importance of public services in attaining sustainability (77) and
on co-operation: "harmony rather than competition, excellence, elitism, separation or
isolation" (178) would be needed.

The Heinrich Böll Foundation (2003b) defines in the *Charter of Civil Rights for a Sustainable Knowledge Society* a sustainable knowledge society as a society based on free
access to knowledge, knowledge as public good owned by all (the Commons), openness
of technical standards and organisation forms, securing privacy, cultural and linguistic diversity, diversity of the media and public opinion, the long-term conservation of
knowledge, bridging the digital divide, freedom of information as a civil right to political activity and transparent administration, and securing freedom in work environment.
This definition takes into account technological, economic, political, and cultural issues,
missing are ecological concerns. A sustainable knowledge society would preserve and
promote human rights, give unhampered and inclusive access to knowledge, provide
means for preserving the natural environment, and provide access to the diverse media
constituting the knowledge of the past (Heinrich Böll Foundation 2003a). The dialectic
of SIS is taken into account by arguing that economisation hinders sustainability: "The
Charter is directed emphatically against the increasing privatisation and commercialisation of knowledge and information. A society, in which the protection of intellectual
property transforms knowledge into a scarce resource, is not sustainable" (Heinrich Böll
Foundation 2003a).

The World Summit on the Information Society (WSIS) Civil Society Plenary (2005) argues
that in the WSIS process civil society interests were not adequately taken into account
(for a critique of WSIS see also Servaes and Carpentier 2006). In its own declaration –
that is very different from the official dualistic WSIS outcome documents –, the WSIS
Civil Society Plenary (2003) argues for an information society that is based on 34 inclusive principles. Among them are the promotion of free software and the establishment
of a public domain of global knowledge that challenges intellectual property. The focus
is on public goods and redistribution. The Plenary stresses that distributive justice is

needed and that economic resources hence need not simply be produced within economic growth models, but need to be redistributed:

> "We aspire to build information and communication societies where development is framed by fundamental human rights and oriented to achieving a more equitable distribution of resources, leading to the elimination of poverty in a way that is non-exploitative and environmentally sustainable".
>
> (WSIS Civil Society Plenary 2003, 3)

Wolf Göhring (1999) speaks of a sustainable productive information society as a society in which humans with web support publicly plan, produce, run, maintain, repair, and take systems out of service in a collaborative way, create networked products and systems so that the free use of machinery, information, resources, free communication, and the free production of goods advances a sustainable society that benefits all. Göhring is concerned with how people have to interact and produce in order to bring about sustainability and corresponding worldviews. His approach is a process- and co-operation-oriented, a SIS would be a truly post-modern society that eliminates instrumental reason.

9.3 Conclusion

The task of this chapter was to comparatively and theoretically grounded discuss notions of sustainability, inclusion, and participation in the information society discourse. A theoretical model of society as a dialectical system was introduced, in which the economic base and the political-cultural superstructure are mutually shaping each other. Based on a distinction between reductionistic, holistic, dualistic, and dialectical worldviews, four different theoretical approaches on defining the SIS were distinguished, which are based on how the relationship between base and superstructure is conceived. Reductionistic approaches see ecological or technological or economic changes as the sole driving forces of a SIS. Projectionistic approaches see superstructures (polity and/or culture) as the determining forces of a SIS. They are the least frequently found approaches in the literature. Dualistic approaches define multiple goals and dimensions of a SIS, but do not consider if these goals are compatible and if and how they are causally linked. Dualistic models are the ones that can be found most frequently in the literature.

As an alternative to these three models, the dialectical notion of the PCSIS was introduced. Co-operation is based on an inclusive logic that establishes social systems, in which all involved actors and groups benefit. The logic of co-operation is the binding force of a progressive society that connects its various dimensions. Participation means the structural enablement, co-operation the intersubjective social process, sustainability the long-term condition and effects of social systems, in which all benefit and have a good life. Abstractly spoken, a PCSIS is a society that guarantees a good life for all. A PCSIS is a society in which knowledge and technology are together with social systems shaped in such ways that humans are included in and self-determine their social systems collectively, interact in mutually benefiting ways, and so bring about a long-term stability that benefits all present and future generations and social groups. As specific qualities of co-operation in a PCSIS, ecological preservation, human-centred technology, socio-economic equity, political freedom, and cultural wisdom are identified and defined.

The task of this chapter was not to quantify to which degree a PCSIS has already been achieved or to suggest indicators of such measurement. This is an empirical research task that needs to be tackled in the future. For doing so, a meta-theory that defines the SIS and provides arguments on which qualities such a society should have and how this could be achieved is needed. Hence, SIS studies need a theoretical and normative grounding. One such approach on socio-theoretical grounding was undertaken in this chapter. Its claim is not to be the only or the ultimate theoretical meta-approach, but the debate thus far lacks a multitude of approaches, and hence this chapter wants to contribute to the discourse on the theoretical groundworks of the debate on SIS.

In this chapter, it was pointed out that the discourse on SIS is dominated by dualistic approaches. In dualistic approaches, various goals are proclaimed, but it is not considered if these goals are compatible. This view has developed into an ideology that stresses various desirable goals such as social cohesion and environmental protection, but at the same time does not question the predominant economic colonisation of society by instrumental reason and the logic of commodities and money capital that has caused a rise in poverty, exclusion, and the income gap during the past decades in Europe, North America, and on the global scale. The problem is that many of the dualistic authors and policy advisors do not realise that capitalistic economic growth is unsustainable as such and inherently produces an antagonism between economic freedom and social equity and that hence systemic alternatives to capitalism must be found in order to truly advance sustainability. The alternative argument made in theús chapter was that late-modern society is characterised by a culminating antagonism between economic growth and social and ecological cohesion, economic freedom (of markets), and social equity. Less profitability and more corporate taxation are needed in order to provide financial means that can be invested in social and ecological sustainability. Economic sustainability hence should not be understood as meaning continuously rising profit rates, but should better be conceived as self-managed ownership, distributive justice, and the advancement of public goods (based on the insight that the commons are produced co-operatively and hence should be owned collectively). The alternative view is shared by a number of institutions such as the Heinrich Böll Foundation, UNESCO, and the World Summit on the Information Society (WSIS) Civil Society Plenary.

In the current discourse, concepts such as sustainability, participation, co-operation, and corporate social responsibility have ideological character, they serve as legitimating predominant capitalist interests that present themselves as open-minded and willing to make some small changes as long as these changes do not question profitability and the capitalist system. Progressive terms that signify inclusion are used for advancing exclusion and capitalist interest (Fuchs 2008b). It is time for a critical alternative to these ideological conceptions. The alternative view of a less-capitalistic or even a non-capitalistic co-operative information society as SIS is generally marginalised and downplayed by dominant actors in discourse. Nonetheless, it is existent and the task for the future is one of academic class struggle that questions ideological dualistic positions and provides arguments that ground the necessity for the transformation towards a PCSIS, that as a precondition is non-capitalistic in character.

References

Archer, Margaret. 2007. Social Integration, System Integration and Global Governance. In *Frontiers of Globalization Research*, ed. Ino Rossi, 221–241. Berlin: Springer.

Bhaskar, Roy. 1993. *Dialectic: The Pulse of Freedom*. London: Verso.

Bourdieu, Pierre. 1986. The (Three) Forms of Capital. In *Handbook of Theory and Research in the Sociology of Education*, ed. John G. Richardson, 241–258. New York: Greenwood Press.

Britton, Eric, ed. 1996. Information Society and Sustainable Development. *Journal of World Transport Policy & Practice* 2 (1).

Carrelli, Claudio, Jan Van Dijk, John Gray, Joan Majo, Robert Pestel, and Franz Josef Radermacher. 2000. Towards a Global Sustainable Information Society. A European Perspective. *Concepts and Transformations* 5 (1): 43–63.

Club of Rome. 2003. *Towards a New Age of Information and Communication for All*. Hamburg: Club of Rome.

Club of Rome and Factor 10 Institute. 2002. *Where Are We Going? Where Do We Want To Be? How Do We Get There? Visions and Roadmaps for Sustainable Development in a Networked Knowledge Society*. Brussels: European Commission.

Commission of the European Communities. 2006. *Communication from the Commission: Building the ERA of Knowledge for Growth*. Brussels: Commission of the European Communities.

Commission of the European Communities. 2005. *i2010. A European Information Society for Growth and Employment.*

Commission of the European Communities. 2002. *eEurope 2005: An Information Society for All. Action Plan to be Presented in View of the Sevilla European Council, 21/22 June 2002.*

Commission of the European Communities. 2000a. *eEurope 2002: An Information Society for All. Action Plan.*

Commission of the European Communities. 2000b. *eEurope: An Information Society for All. Communication on a Commission Initiative for the Special European Council of Lisbon, 23 and 24 March 2000.*

Davies, James B., Susanna Sandstrom, Anthony Shorrocks, and Edward N. Wolff. 2006. *The World Distribution of Household Wealth.* https://www.gtap.agecon.purdue.edu/resources/download/3063.pdf (accessed on 26 March 2021).

DEMO-net. 2006. *Demo-net: The eParticipation Workshop: Mapping eParticipation.* White Papers.

European Commission. 1998. *1998 Status Report: Toward A Sustainable Information Society.* Brussels: Commission of the European Communities.

Fuchs, Christian. 2008a. *Deconstructive Class Analysis: Theoretical Foundations and Empirical Examples for the Analysis of Richness and the Class Analysis of the Media and the Culture Industry.* ICT&S Research Paper No. 4. Salzburg: ICT&S Center.

Fuchs, Christian. 2008b. *Internet and Society: Social Theory in the Information Age.* New York: Routledge.

Fuchs, Christian. 2008c. The Implications of New Information and Communication Technologies for Sustainability. *Environment, Development and Sustainability* 10 (3): 291–310.

Giddens, Anthony. 1984. *The Constitution of Society. Outline of the Theory of Structuration.* Cambridge: Polity Pres.

Göhring, Wolf. 1999. *The Productive Information Society: A Basis for Sustainability.* Sankt Augustin: GMD – Forschungszentrum Informationstechnik GmbH.

Habermas, Jürgen. 1981. *Theorie des kommunikativen Handelns.* 2 vols. Frankfurt am Main: Suhrkamp.

Hall, Stuart. 1983. The Problem of Ideology: Marxism without Guarantees. In *Marx: A Hundred Years On*, ed. Betty Metthews, 57–84. London: Lawrence & Wishart.

Harvey, David. 2005. *A Brief History of Neoliberalism.* Oxford: Oxford University Press.

Hegel, Georg Wilhelm Friedrich. 1830. *Enzyklopädie der philosophischen Wissenschaften im Grundrisse. Erster Teil: Die Wissenschaft der Logik. Werke, Band 8.* Frankfurt am Main: Suhrkamp.

Heinrich Böll Foundation. 2003a. *Charter of Civil Rights for a Sustainable Knowledge Society.* Version 2.0, May 2003. http://www.worldsummit2003.de/en/web/375.htm (accessed on 19 September 2007).

Heinrich Böll Foundation. 2003b. *Towards A "Charter of Human Rights for Sustainable Knowledge Societies".* https://www.itu.int/dms_pub/itu-s/md/03/wsispc2/c/S03-WSISPC2-C-0065!!PDF-E.pdf (accessed on 26 March 2021).

Hilty, Lorenz. 2000. Towards a Sustainable Information Society. *Informatique* August 2000: 2–9.

Hofkirchner, Wolfgang. 2002. *Projekt Eine Welt: Kognition, Kommunikation, Kooperation.* Münster. LIT.

Holz, Hans Heinz. 2005. *Weltentwurf und Reflexion: Versuch einer Grundlegung der Dialektik.* Stuttgart: J.B. Metzler.

Information Society Forum. 2000. *A European Way for the Information Society.* http://www.poptel. org.uk/nuj/mike/isf/ew.html (accessed on 19 September 2007).

Information Society Forum. 1998. Forum Info 2000: Challenges 2025 – On the Way to a Sustainable World-Wide Information Society. In *1998 Status Report: Toward A Sustainable Information Society*, ed. European Commission, 91–97. Brussels: Commission of the European Communities.

Kant, Immanuel. 1998. *Groundwork of the Metaphysics of Morals.* New York: Cambridge University Press.

Lisbon European Council. 2000. *Presidency Conclusions, 23 and 24 March 2000.*

López-Ospina, Gustavo. 2003. *Planetary Sustainability in the Age of the Information and Knowledge Society. For a Sustainable World and Future. Working Toward 2015.* Paris: UNESCO.

Luhmann, Niklas. 1984. *Soziale Systeme.* Frankfurt am Main: Suhrkamp.

MacIntosh, Ann. 2006. eParticipation in Policy-Making: the Research and the Challenges. In *Exploiting the Knowledge Economy: Issues, Applications, Case Studies,* ed. Paul Cunningham and Miriam Cunnigham, Amsterdam: IOS Press 364–369.

Macintosh, Ann. 2004. Characterizing e-participation in Policy-Making. In *Proceedings of the Thirty-Seventh Annual Hawaii InternationalSciences (HICSS-37)*, Big Island, Hawaii, January 5–8.

Macpherson, C.B. 1973. *Democratic Theory: Essays in Retrieval.* Oxford: Oxford University Press.

Milanovic, Branko. 2007a. *An Even Higher Global Inequality than Previously Thought: A Note on Global Inequality Calculations Using the 2005 ICP Results.* http://ssrn.com/abstract=1081970 (accessed on 26 March 2021).

Milanovic, Branko. 2007b. Globalization and Inequality. In *Global Inequalities*, ed. David Held and Ayse Kaya, 26–49. Cambridge: Polity.

Milanovic, Branko. 2002. True World Income Distribution, 1998 and 1993: First Calculation Based on Household Surveys Alone. *Economic Journal* 112: 51–92.

O'Donnell, Susan. 2001. *Towards an Inclusive Information Society in Europe: The Role of Voluntary Organisations. IST Study Report Research Programme European Commission.* Dublin: Models Research.

O'Donnell, Susan, Helen McQuillan, and Anna Malina. 2003. *eInclusion: Expanding the Information Society in Ireland.* Dublin: Itech Research.

Radermacher, Franz Josef. 2004. *Global Marshall Plan: For a Worldwide Eco-Social Market Economy.* Hamburg: Global Marshall Plan Initiative.

Schauer, Thomas. 2003. *The Sustainable Information Society: Vision and Risks.* Ulm: Universitätsverlag Ulm.

Serageldin, Ismail. 1995. The Human Face of the Urban Environment. In *Proceedings of the Second Annual World Bank Conference on Environmentally Sustainable Development: The Human Face of the Urban Environment, Washington, D.C., September 19–21, 1994*, ed. Ismail Serageldin, Michael A. Cohen, and K.C. Sivaramakrishnan, 16–20. Washington, DC: World Bank.

Servaes, Jan and Nico Carpentier, eds. 2006. *Towards a Sustainable Information Society. Deconstructing WSIS.* Bristol: Intellect.

Stiglitz, Joseph E. 2003. *Globalization and its Discontents.* New York: Norton.

UNHDR. 2008. *Human Development Report 2007/2008: Fighting Climate Change.* New York: UNDP. https://www.undp.org/publications/human-development-report-2007/2008

Williams, Raymond. 2001. *The Raymond Williams Reader,* ed. John Higgins. Malden, MA: Blackwell.

World Commission on Environment and Development (WCED). 1987. *Our Common Future.* Oxford: Oxford University Press.

World Summit on Sustainable Development. 2002. *The Jo'burg Memo. Fairness in a Fragile World.* Berlin: Heinrich Böll Foundation.

World Summit on the Information Society (WSIS). 2005. *Outcome Documents: Geneva 2003 – Tunis 2005.* Geneva: ITU.

World Summit on the Information Society (WSIS) Civil Society Plenary. 2005. *Much More Could Have Been Achieved: Civil Society Statement on WSIS.* https://waccglobal.org/wp-content/uploads/2020/07/Much-more-could-have-been-achieved.pdf (accessed on 26 March 2021).

World Summit on the Information Society (WSIS) Civil Society Plenary. 2003. *Civil Society Declaration to the World Summit on the Information Society: Shaping Information Societies for Human Needs.* https://www.itu.int/net/wsis/docs/geneva/civil-society-declaration.pdf (accessed on 26 March 2021).

Part III

Conclusion

Chapter Ten
The Digital Commons and the Digital Public Sphere: How to Advance Digital Democracy Today

10.1 Introduction

In the past 15 years, the notions of big data and social media have become prevalent in everyday life. Associated with it, we have experienced the rise of platforms such as Google, YouTube, Facebook, TikTok, Amazon, Twitter, Apple, Baidu, Instagram, WhatsApp, WeChat, Alibaba, Spotify, or Netflix. These platforms gather lots of personal user data and provide services such as search engines, video platforms, social networks, online shop microblogs, photo-sharing platforms, messenger apps, or music and film streaming.

This chapter asks: what are the democratic potentials of the digital commons and the digital public sphere? First, the chapter identifies ten problems of digital capitalism. Second, it engages with the notion of the digital public sphere. Third, it outlines the concept of the digital commons. Fourth, some conclusions are drawn and ten suggestions for advancing digital democracy are presented.

10.2 Digital Capitalism

Capitalism is a type of society that is based on the logic of the accumulation of power (Fuchs 2020a). Money capital is a particular and important form of power that is accumulated in capitalist society. But the logic of accumulation also shapes politics and culture. Politics in capitalist society is the sphere of the accumulation of decision-power. Culture in capitalist society is the sphere of the accumulation of reputation. Inequality and injustices are the consequences of the logic of accumulation: the capitalist economy

DOI: 10.4324/9781003279488-13

is shaped by the exploitation of labour and the asymmetric distribution of wealth; the capitalist political system is shaped by domination and asymmetrical influence; the capitalist cultural system is shaped by ideology, malrecognition, and disrespect.

Digital capitalism is not a new phase of capitalist development, but rather a dimension of the organisation of capitalism that is shaped by digital mediation. In digital capitalism, social processes, such as the accumulation of power, capital accumulation, class struggles, political struggles, hegemony, ideology, commodification, or globalisation, are mediated by digital technologies, digital information, and digital communication. Transnational digital and communication corporations play an important role in digital capitalism.

Twenty-one of the world's largest 100 transnational corporations operate in the communication, media, and digital industry (Table 10.1). Subsectors of the capitalist communication, media, and digital industry include, for example, advertising, broadcast networks, cloud storage, communication/digital networks, digital games, digital hardware, digital services and platforms, leisure and live entertainment culture, online shopping, online streaming, or software. The total profits of the dominant 21 communication/digital/media corporations amounted in the financial year 2019 to USS$2.5 trillion, which made up 3% of the global 2019 Gross Domestic Product.[1] That just 21 companies control 3% of the world's financial wealth produced during one year shows the large power of capitalist companies, including digital and communication corporations.

TABLE 10.1 The domination transnational communication and digital corporations, data sources: Forbes 2000 List (year 2020), https://www.forbes.com/global2000, accessed on 7 October 2020

Rank	Company	Country of headquarter	Commodities	Annual profits (US$ bn)	Annual revenues (US$ bn)
9	Apple	USA	Digital hardware, software, digital services, online streaming, cloud storage	267.7	57.2
11	AT&T	USA	Communication/digital networks, broadcasting networks	179.2	14.4
13	Alphabet	USA	Digital advertising, digital services and software (e.g. Google, YouTube, Android OS, Chrome)	166.3	34.5
13	Microsoft	USA	Software (e.g. Windows, Office, Skype), hardware (Xbox)	138.6	46.3
16	Samsung Electronics	South Korea	Digital hardware	197.6	18.4

TABLE 10.1 (Cont.)

Rank	Company	Country of headquarter	Commodities	Annual profits (US$ bn)	Annual revenues (US$ bn)
20	Verizon Communications	USA	Communication/digital networks	131.4	18.4
22	Amazon	USA	Online shopping platform, online streaming of digital content, cloud storage	296.3	10.6
27	Comcast	USA	Communication/digital networks, broadcast networks	108.7	11.7
28	China Mobile	China	Communication/digital networks	108.1	15.5
31	Alibaba	China	Online shopping platforms, online payment, online entertainment streaming	70.6	24.7
36	Walt Disney	USA	Entertainment content, broadcast networks, digital entertainment and streaming platforms, leisure parks, merchandising	74.8	10.4
38	Intel	USA	Digital hardware	75.7	22.7
39	Facebook	USA	Digital advertising, digital services (Facebook, Instagram, WhatsApp)	73.4	21
43	Nippon Telegraph & Tel	Japan	Communication/digital networks	109.6	7.9
50	Tencent	China	Digital advertising, digital services (QQ, WeChat), digital games, music and video streaming,	54.6	13.5
51	IBM	USA	Digital hardware, software, cloud computing	76.5	9.0
58	Sony	Japan	Hardware, video games, entertainment content	79.2	6
66	SoftBank	Japan	Communication/digital networks, digital hardware	87.4	2.9
69	Deutsche Telekom	Germany	Communication/digital networks	90.1	4,3
82	Cisco Systems	USA	Digital hardware, software	51.6	11.1
94	Oracle	USA	Digital hardware, software	39.8	10.8
			Total	2,477.2	367.0

Digital capitalism has been shaped by ten major societal problems (Fuchs 2021, especially Chapter 14):

1) Communication and digital capital exploits communication and digital labour and has resulted in the tendency of capitalist **monopolies** in the communication and digital industry.

2) Dominant Internet culture is shaped by a competitive, **individualistic digital culture** that is me-centred and focused on the accumulation and asymmetric distribution of online attention, influence, reputation, visibility, and voice.

3) Communication/digital corporations and state apparatuses have created a **surveillance-industrial complex**.

4) Capitalist social media are **anti-social media** that have helped advancing anti-democratic potentials, digital authoritarianism, digital racism, digital nationalism, and digital fascism.

5) In **algorithmic politics**, algorithms create online content and attention and it becomes difficult for humans to discern which online activities are human and which ones are machinic.

6) In the online world, there are **fragmented digital public spheres** where we find filter bubbles

7) The digital culture industry has created **digital ideology**, ideologies about the digital, and ideology disseminated via digital networks. Digital advertising and tabloid content dominate the online world. Many digital platforms are **digital tabloids**.

8) **Influencer capitalism** dominates social media and has created asymmetric attention, reputation, and visibility on the Internet as well as an ideological culture dominated by shopping and advertising. Advertising is increasingly hidden and presented as regular content ("branded content").

> "Influencer capitalism is not a type of capitalism but an ideology that claims that by being active on social media platforms such as Instagram, Snapchat, and YouTube there are great opportunities for becoming wealthy and famous. Influencer capitalism is the dream, fantasy, and desire of users to become celebrities that accumulate a wealth of social relations, money, influence, likes, positive comments, etc. Influencer capitalism is the online manifestation of the American Dream's ideological claim that in capitalism everyone has an equal opportunity to make a career, from a dishwasher to a billionaire, by having a good idea and believing in themselves".
>
> (Fuchs 2021, 175)

9) The high amount of online information flows processed at high speed has resulted in **digital acceleration**. There is a lack of time and space for sustained political debate.

10) On social media, one frequently encounters **fake/false news and post-factual politics** that deny facts and are led by emotionalisation, tabloidisation, and ideology.

The combined consequences of these ten developments are that democracy is under threat and we have experienced the rise of authoritarian capitalism where far-right demagogues dominate politics (Fuchs 2018a, 2020b). Digitalisation is not the cause of these developments, but has rather mediated the antagonism between neoliberal capitalism and rising social inequalities. The commodification, privatisation, commercialisation, and individualisation of (almost) everything have turned against liberalism' civic values and political freedoms, which has given rise to new nationalist, racist, xenophobic, authoritarian, and fascist forces in society.

The question arises what the alternatives are to digital capitalism and digital authoritarianism. Are the digital public sphere and the digital commons such alternatives?

10.3 The Digital Public Sphere

Political communication is an important and indispensable aspect of the political system in all models of democracy. In general terms, it can be said that the public is a central mechanism of the political system. By "public" we generally understand goods and spaces that are "open to all" (Habermas 1989, 1). One speaks, for example, of public education, public buildings, public parks, public squares, public assemblies, public demonstrations, public opinion, public media, etc. Public goods and facilities are not reserved for a clique or a club of the privileged, but are intended for the general public, i.e. all members of society.

The public sphere is a sphere of public political communication that mediates between the other subsystems of society, namely, the economy, politics, culture, and private life. The ideal type of the public sphere is a realm of society that organises "critical publicity" (Habermas 1989, 237) and "critical public debate" (Habermas 1989, 52). The public sphere mediatises political communication. It is a mediatising space of political communication in which citizens meet, who inform themselves about life in society and communicate politically. The public sphere is a space where political opinions are formed. Public communication is an important aspect of the existence of humans as social beings and of society. In modern society, the media system is the most important organised form of public communication. In the media system, media actors produce public information.

There is a number of criticisms of Habermas's concept of the public sphere, mainly from the field of postmodern studies. The present author has in other places criticised the dismissal of Habermas and the public sphere concept and argues that Habermas's concept is useful in and can be updated to the digital age (see Fuchs 2014b).

The digital public sphere is not a separate sphere of society, but a dimension and aspect of the public sphere in societies where digital information and digital communication are prevalent. The digital public sphere means the publishing of information, critical publicity, and critical public debate mediated by digital information and communication technologies. Not all information and communication via the Internet, mobile phones, and tablets are part of the digital public sphere. When processes of commodification and capitalisation (the logic of economic accumulation), domination (the logic of political accumulation), and ideology (the logic of political accumulation) shape digital practices, the latter do not form a public sphere. The digital public sphere has then, as Habermas (1989) argues, been colonised and feudalised. We can then speak of an alienated digital sphere and alienated communication but not of a digital public sphere. The ten processes outlined in the previous section are manifestations of digital alienation, digital colonisation, and digital feudalisation.

The mentioned ten tendencies lead overall to a digital sphere that is characterised and divided by economic, political, and cultural power asymmetries. The logics of accumulation, advertising, monopolisation, commercialisation, commodification, acceleration, individualism, fragmentation, automation of human activity, surveillance, and ideologisation turn the digital public sphere into a colonised and feudalised sphere, a pseudo-digital public sphere that is public in appearance only. In digital capitalism, commercial culture dominates the Internet and social media. Platforms are largely owned by profit-oriented corporations. Public service media operate on the basis of a different logic. However, the idea of a public Internet has not yet been able to establish itself and sounds strange to most ears, as there are hardly any alternatives to the commercial Internet today.

Public service media are media of, in, and operating through the public sphere. The communication scholar Slavko Splichal (2007, 255) gives a precise definition of public service media:

> "In normative terms, public service media must be a service *of* the public, *by* the public, and *for* the public. It is a service *of* the public because it is financed by it and should be owned by it. It ought to be a service *by* the public – not

only financed and controlled, but also produced by it. It must be a service *for* the public – but also for the government and other powers acting in the public sphere. In sum, public service media ought to become 'a cornerstone of democracy'".

The means of production of public service media are publicly owned. The production and circulation of content are based on a non-profit-making logic and the public service media remit. Access is universal, as all citizens are given easy access to the content and technologies of public service media. In political terms, public service media offer diverse and inclusive content that promotes political understanding and discourse. In cultural terms, they offer educational content that contributes to the cultural development of individuals and society. Public service media have a special, legally defined remit, namely, that they have to produce and provide content and services that help to advance democracy, education, and culture. In debates, public service media such as the BBC are often incorrectly presented as state-media or state-controlled media. True public service media are legally enabled by the state (licence fee funding, public service remit), but not controlled by the state. Public service media are independent media organisations that are enabled by state laws.

Due to the special qualities of public service media, they can also make a particularly valuable democratic and educational contribution to a democratic online public sphere and digital democracy if they are given the necessary material and legal means to do so.

Life in modern society has increasingly been accelerated, which includes the acceleration of the economy, political decision-making, lifestyles, and experiences (Fuchs 2014a; Rosa 2013). The logic of accumulation is the driving force of acceleration (see Fuchs, 2014a). As a consequence, the speed of social relations has been increased, especially since the rise of neoliberal capitalism. In the realm of the media, the acceleration of information flows has been an aspect of the tabloidisation of media and communication that in turn is an aspect of the commercialisation, monopolisation, and commodification of the media.

The predominant media are high-speed spectacles that are superficial and characterised by a lack of time provided for debate. They erode the public sphere and the culture of political discussion. They leave no time or space to citizens for grasping the complexity of society and for developing arguments. What we need today is the decommodification and deceleration of the media. We need slow media (see Fuchs 2021; Köhler, David and Blumtritt, 2010; Rauch 2018).

The Digital Public Sphere

What is slow media?

> "Slow media takes the speed out of information, news, and political communication by reducing the amount of information and communication flows. Users engage more deeply with each other and with content. Slow media does not distract users with advertisements, it is not based on user surveillance, and it is not undertaken to yield profit. It is not simply a different form of media consumption, but an alternative way of organising and doing media – a space for reflection and rational political debate".
>
> (Fuchs 2021, 363)

Slow media and slow political communication are not new. Club 2 in Austria and After Dark in the UK are prototypical examples. The journalists Kuno Knöbl and Franz Kreuzer designed the Club 2 concept for the Austrian Broadcasting Corporation (ORF). It was a discussion programme which was usually broadcast on Tuesday and Thursday. The first programme was broadcast on 5 October 1976, the last on 28 February 1995. 1,400 programmes were broadcast on ORF (Der Standard 2001). Club 2 had a new edition that was broadcast from 2007 to 2012. However, a different concept was used that did not adhere to the original principles.

In the UK, the media production company Open Media created a similar format based on Club 2 under the name After Dark. After Dark was broadcast once a week on Channel 4 between 1987 and 1991 and occasionally thereafter. In 2003, After Dark was shown for a short time on BBC.

The producer of After Dark Sebastian Cody describes the Club 2/After Dark concept as follows:

> "Namely, the number of participants in these intimate debates (always conducted in agreeable surroundings and without an audience) was never less than four, never more than eight (like, as it happens, group therapy); the discussion should be hosted by a non-expert, whose job rotates, thus eliminating the cult of personality otherwise attaching to presenters; the participants should be a diverse assortment, all directly involved in the subject under discussion that week; and, most importantly, the programme was to be transmitted live and be open-ended. The conversation finishes when the guests decide, not when TV people make them stop".
>
> (Cody 2008)

The concept of Club 2 sounds rather unusual to many people today, as we are so used to formats with short duration, high speed, and the lack of time in the media and our daily lives. Open, uncensored, controversial, live discussions that appeal to the viewer and the audience are different from accelerated media in terms of space and time: Club 2 was a public space where guests met and discussed with each other in an atmosphere that offered unlimited time, which was experienced publicly and during which a socially important topic was discussed. Club 2 was a democratic public space in public broadcasting.

Space and time are two important dimensions of the public political economy. However, a social space that offers enough time for discussion is not yet a guarantee for a committed, critical and dialectical discussion that transcends one-dimensionality, penetrates into the depths of a topic, and highlights the similarities and differences of different positions. Public space and time must be organised and managed in an intelligent way, so that the right people participate, the atmosphere is appropriate, the right discussion questions are asked, and it is ensured that all guests have their say, listen to each other and that the discussion can proceed undisturbed, etc. Unlimited space, a dialectically controversial and intellectually challenging space, and intelligent organisation are three important aspects of publicity. These are preconditions of slow media, non-commercial media, decolonised media, and media of public interest.

We need slow media. Online and offline. Let's decelerate the media and create slow media 2.0. Is a new version of Club 2 (Club 2.0) as part of the digital public sphere possible today?

Club 2.0 is an example of a public service Internet platform that helps advancing democratic communication and the digital public sphere. In Club 2.0, the traditional principles of Club 2 are practiced and updated (see Figure 10.1). There is a controversial live studio debate without time limit. It is broadcast on television and also on a public service video platform, a public service version of YouTube. Social media enable user-generated content and online debate. An updated version of Club 2 should make use of the affordances of digital media: in Club 2.0, users can upload discussion inputs and discuss in text- and video-based formats on the public service YouTube channel that accompanies the Club 2.0 television broadcasts. At certain points of time of the TV debate, single user-generated video discussion inputs are selected and broadcast as part of the television discussion so that they inform the studio debate.

Club 2.0 is an expression of digital democracy and the digital public sphere. It manifests a combination of elements of deliberative and participatory democracy.

The Digital Public Sphere

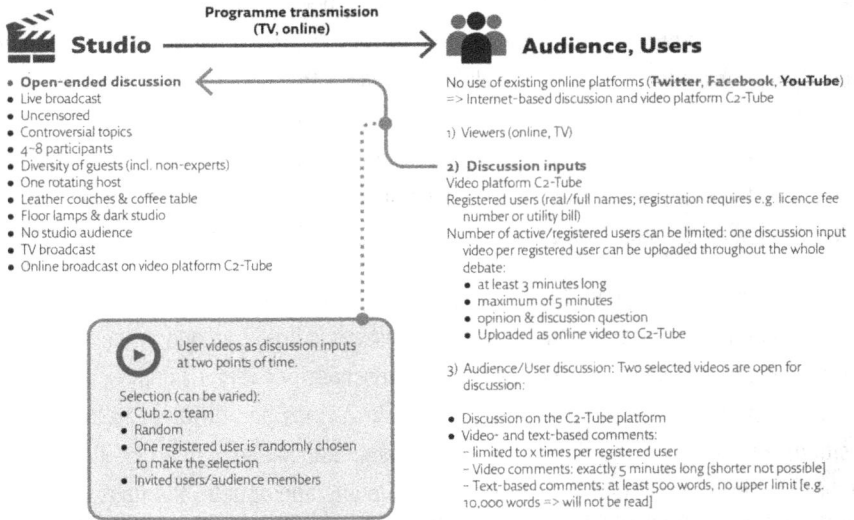

FIGURE 10.1   Club 2.0, first published as Creative Commons in Fuchs (2018b, 74).

Deliberative democracy creates "institutional designs of modern democracy" that are based on the "principle of reciprocity" (Held 2006, 233). It enables "social encounters which take account of the point of view of others – the moral point of view" (Held 2006, 233), places "greater emphasis upon those settings and the procedures of preference formation and learning within politics and civil society" (Held 2006, 233) and the "giving of defensible reasons, explanations and accounts for public decisions" (Held 2006, 237). It stresses the communicative dimension of democracy and that "[c]ritical reflection must link up with public debate and deliberative politics" (Held 2006, 241). Club 2.0 is a communicative mechanism that allows more reflection, explanation, and debate in politics and can thereby strengthen deliberative aspects of democracy. Participatory democracy stresses "[d]irect participation of citizens in the regulation of the key institutions of society, including the workplace and local community" (215). Club 2.0 increases the participation of citizens in culture and political debate. It is a participatory aspect of culture and politics. Club 2.0 offers space and time for controversial political communication and enables citizens to participate in the discussion collectively and individually through videos and commentaries. Club 2.0 brings together the communicative aspect of deliberative democracy and the participatory idea of grassroots democracy. The digital public sphere and public service Internet platforms are social phenomena that are opposed to and challenge digital capitalism and the capitalist Internet.

10.4 The Digital Commons

Elinor Ostrom, the winner of the Nobel Memorial Prize in Economic Sciences 2009, defines the commons with a theory of economic goods that is based on the two features of exclusion and subtractability. For Ostrom, common-pool resources have high subtractability and low exclusivity (Hess and Ostrom 2007, 9). An example is a library with open accessibility: it has a low entry barrier (low exclusivity) and high subtractability (it becomes unusable if too many humans use it at once). In contrast, she argues that a sunset is a public good because it is non-rivalrous in consumption (low subtractability) and it is difficult to exclude someone from sunshine and watching the sunset (low exclusivity).

The problem of Ostrom's concept is that it neglects political economy, i.e. the concept of common ownership. Ostrom thereby de-politicises the concept of the commons. Yochai Benkler argues that influenced by Ostrom's works, "a more narrowly defined literature developed" (Benkler 2006, 480). Benkler argues for "an entirely different theory of the commons" (Benkler 2013, 1510). He defines the commons in the following way:

> "Commons are an alternative form of institutional space, where human agents can act free of the particular constraints required for markets, and where they have some degree of confidence that the resources they need for their plans will be available to them. Both freedom of action and security of resource availability are achieved in very different patterns than they are in property-based markets".
>
> (Benkler 2006, 144)

For Benkler, the commons are non-market and non-profit-based resources that are available to everyone (for another definition, see Bauwens, Kostakis and Pazaitis 2019, 3). Slavoj Žižek (2010, 212–213) identifies three forms of the commons:

- the cultural commons: language, means of communication, education, infrastructures),
- the commons of external nature: the natural environment,
- the commons of internal nature: the human being.

Michael Hardt and Antonio Negri (2017 166) identify two basic forms of the commons, namely the social and the natural commons. They subdivide these two types into five kinds of the commons:

- the natural commons: ecosystems, the earth;
- the social commons 1: codes, ideas, images, cultural products;

The Digital Commons

- the social commons 2: physical products commonly produced by co-operative work;
- the social commons 3: rural and metropolitan spaces where human communicate, co-operate and interact culturally;
- the social commons 4: institutions that provide health care, education, housing, and welfare for all.

Already Karl Marx stressed that there are resources in society that are produced in a collective and co-operative manner. He argues that universal work creates common goods that are "brought about partly by the cooperation of men now living, but partly also by building on earlier work" and involve the "direct cooperation of individuals" (Marx 1894, 199). Commons are resources that are collectively owned and that are produced in a co-operative manner. The natural commons are resources produced by nature that are required for all humans to survive. It includes the earth and the universe as the natural habitat of humans. Nature constantly produces and reproduces itself. It is a self-organising system. Nature as such is by its own nature a common good because when it produces itself it is available to everyone.

But historically, capital expropriated and enclosed parts of nature so that they became private property. In the middle ages, humans used the land, the forests, the fields, the meadows, etc. as common goods. The formation of capitalism involved what Marx (1867, part 8) terms original primitive accumulation, the violent transformation of humans into wage labourers. One measure was the legal enclosure of the natural commons so that land became private property. Peasants were driven from the land and henceforth had to earn a living as wage workers.

Socially produced goods are commons when they are collectively owned and co-operatively produced. There is an important moral-political principle underlying Marx's and Engels's thought and politics: those who produce the goods should collectively own them. For Marx and Engels, the central characteristic of a communist society is that there is common ownership of the means of production by the workers:

"In this sense, the theory of the Communists may be summed up in the single sentence: Abolition of private property. [...] When, therefore, capital is converted into common property, into the property of all members of society, personal property is not thereby transformed into social property. It is only the social character of the property that is changed. It loses its class character".

(Marx and Engels 1848, 498 and 499)

TABLE 10.2 Four types and dimensions of the commons

Sphere of society	Type of the commons	Meaning of the commons
Nature	Natural commons: environmental sustainability	Common access to natural resources for everyone, common use of natural resources in environmentally sustainable manners
Economy	Economic commons: socialism	Common ownership of the means of production, wealth for and self-realisation of all humans
Politics	Political commons:participatory democracy	Humans who are affected by certain phenomena can take collective decisions about these affairs, basic political rights are guaranteed for all and commonly respect
Culture	Cultural commons:culture of friendship	All humans are respected and they are able to understand each other and live together through common practices in everyday life so that friendships and a unity of diversity of lifestyles, identities, and communities are possible

According to Marx and Engels, the commons are not goods that have certain features as assumed in the theory of economic goods. Rather, any good can be transformed into collective ownership. They argue that the means of production should be common goods. A key feature of neoliberal capitalism has been the transformation of common goods into private property and commodities as part of the process that David Harvey (2005, 165–172) calls the commodification of everything. Commodification is an economic process that destroys the material foundations of the commons. It turns something that is available and accessible to all and benefiting all into a private property controlled traded on markets. Utman (2020) points out that in the realm of communication, neoliberal capitalism has resulted in the expropriation of voice as a common resource and practice and has thereby undermined democracy. Based on a model of society, Table 10.2 identifies four types and dimensions of the commons.

Euler (2018) stresses that there is a structural and a practice dimension of the common. "Commoning can be considered the social practices that make commons what they are. [...] commons is the social form of (tangible and/or intangible) matter that is determined by commoning" (Euler 2018, 12). In capitalism, the commons can only exist as seeds of a commons society (Euler 2018, 12). Antonis Broumas (2020, 11–14) points out the commons' dialectic of resource and community. Given that the commons are not just resources, but common resources embedded into practices of commoning and commons communities (Papadimitropoulos 2020, Chapter 1), there is a "distinctive communicative element" of commoning (Utman 2020, 158).

The digital commons are digital resources that are commonly controlled by humans. Table 10.3 presents four types and dimensions of the digital commons

TABLE 10.3 Four types and dimension of the digital commons

Sphere of society	Type of the digital commons	Meaning of the digital commons
Nature	Natural digital commons: digital environmental sustainability	Common control of the mines where natural resources are extracted that form the physical foundations of digital technologies, sustainable environmental impacts of digital technologies that guarantee the common survival of nature, humans, and society (e.g. green computing)
Economy	Economic digital commons: digital socialism	Common ownership of the digital means of production, the use of digital technologies for the advancement of wealth for and self-realisation of all humans
Politics	Political digital common: participatory digital democracy	Collective governance of decisions about the use of digital resource
Culture	Cultural digital commons: digital friendships	Unity in diversity and common recognition and respect of everyone in digitally mediated communities so that friendships are enabled

At the level of digital infrastructures, community networks run as co-operatives are examples of a digital commons projects. At the level of software and digital content, free software and non-commercial Creative Commons licences are examples of digital commons projects. Free, libre, and open-source software (FLOSS) has postcapitalist potentials, but has also in various forms been subsumed under capital (see Berlinguer 2020; Birkinbine 2020). At the level of digital platforms, platform co-operatives are examples of digital commons projects. Platform co-operatives are not-for-profit Internet platforms that are collectively owned and governed by the digital workers who produce the resources that underpin these platforms (see Sandoval 2020; Scholz 2016, 2017; Scholz and Schneider 2016). Examples of platform co-ops are the music platform Resonate (an alternative to Spotify, https://resonate.is/), Fairbnb (an alternative to Airbnb, https://fairbnb.coop/), Taxiapp (an alternative to Uber), the photography and video platform Stocksy (an alternative to Shutterstock and iStockPhoto, https://www.stocksy.com), or the collaboration platform Loomio (https://www.loomio.org/).

Co-operatives in the realm of the digital economy advance the economic commons and the political commons because they are non-profit organisations that are collectively governed and controlled. They are not the only digital commons project. For example, public service Internet projects advance the economic commons as they are owned by the public and the political and cultural commons as they are based on public service remits. Digital commons projects do not automatically advance all levels of the commons. For example, community networks do not necessarily reduce e-waste and energy consumption (environmental sustainability). Some aspects of the commons are per definition covered by digital commons projects, whereas others are only achieved by active commitment beyond the foundation of particular projects.

Open access publishing has emerged as a response to the monopoly practices of capitalist publishers. Open access journals and publishing houses very frequently use Creative Commons licences, which make the content of published works a digital common in the sense that it is a common good that can be accessed by anyone and isn't the exclusive private property of someone but a form of knowledge that is provided to humanity as a gratis resource. On 8 October 2020, 15,273 open access journals were listed in the Directory of Open Access Journals (https://doaj.org).

But open access is not automatically a true digital common that is a manifestation of all four forms of the digital commons identified in Table 10.3. Capitalist open-access publishers have subsumed open access under capital (see Knoche 2020). These are for-profit open access publishers that accumulate capital. The capital accumulation strategy they employ most frequently is that they charge high fees to authors that do not just cover production costs but also yield profits that are privately owned. In capitalist open access, digital content is de-commodified, i.e. the articles and books are published as Creative Commons, but the principles of capital accumulation, commodification, valorisation, and profitability are not given up, but transformed. The opportunity to get published is commodified while the published content is a commons. The digital commons thereby are subsumed under and colonised by digital capital. Capitalist open access is a digital capitalism of the commons. "In the Corporate Open Access Model, companies, organizations or networks publish material online in a digital version, do so free of charge for the readers, but derive monetary profits with strategies such as charging authors or selling advertising space" (Fuchs and Sandoval 2013, 438).

Commons licences should focus on advancing not-for-profit projects that are seeds of postcapitalism. Commons licences such as Creative Commons are not automatically critical of capitalism, some of them are compatible with, subsumed under and supportive of capitalism. In contrast, diamond open access projects are true digital commons projects that have a non-capitalist character. Diamond open access is

> "a form of non-profit academic publishing that makes academic knowledge a common good, reclaims the common character of the academic system and entails the possibility of fostering job security by creating public service publishing jobs. [...] *In the Diamond Open Access Model, not-for-profit, non-commercial*

organizations, associations or networks publish material that is made available online in digital format, is free of charge for readers and authors and does not allow commercial and for-profit re-use".

(Fuchs and Sandoval 2013, 428, 438)

Radical Open Access is a network of diamond open access projects that

"promote a progressive vision for open publishing in the humanities and social sciences. [...] We also share a willingness to subject some of our most established scholarly communication practices to creative critique, together with the institutions that sustain them (the university, the library, the publishing house and so on). The collective thus offers a radical 'alternative' to the conservative versions of open access that are currently being put forward by commercially-oriented presses, funders and policy makers. [...] By showcasing the wide variety of non-commercial, not-for-profit and/or commons-based models for the creation and dissemination of academic knowledge that are currently available, we endeavour to help generate and sustain diversity within the publishing ecology".

(https://radicaloa.disruptivemedia.org.uk/about/)

The question that remains to be answered is if and how the public service Internet projects and the digital commons can contribute to advancing digital democracy. The conclusion addresses this issue.

10.5 Conclusion: Advancing Digital Democracy

The digital public sphere is a dimension of the public sphere, where published knowledge takes on digital formats and informs critical public debate. For Habermas, the public sphere has a democratic, non-capitalist, and unideological character. Therefore, not all digital knowledge and not all digital communication are part of the public sphere. Public service media are media that are publicly owned by independent not-for-profit organisations that are not controlled but enabled by the state and operate basically on the public service remit to provide content and services that advance democratic communication, education, and culture. Public service Internet platforms are Internet platforms owned, operated, and maintained by public service media. Just like public service media, public service Internet platforms are media of, in, and operated through the public sphere. Digital civil society projects such as diamond open access platforms and platform cooperatives are not-for-profit digital projects that are commonly owned and governed by the workers who produce the resources underpinning these projects. Public service Internet platforms and digital civil society/community media platforms are both part of the public sphere and the digital public sphere. Their main difference is that the organisation operating, controlling, and owning the platform is in the first case a public service media organisation and a civil society group or community in the second case. Public service Internet platforms operate closer to the state than platform co-operatives and other digital civil society projects. Public service Internet platforms are, however, not controlled, but rather enabled by the state.

Table 10.4 outlines some foundations of three political economies of digital platforms. Public service Internet platforms and civil society Internet platforms are the two types of digital platforms that operate on non-capitalist principles and thereby negative the political economy of digital capitalism. They operate in the digital public sphere. In contrast, capitalist digital platforms colonise, feudalise, alienate, and destroy the digital public sphere. Public service Internet platforms and civil society Internet platforms are excellent foundations for advancing the digital commons, i.e. digital environmental sustainability (natural digital commons), digital socialism (economic digital commons), participatory digital democracy (political digital commons), and digital friendships (the cultural digital commons). Creating such non-capitalist digital platforms is not a sufficient condition for the advancement of the digital commons, but a good foundation that has a better likelihood and chances to advance digital democracy, digital equality, and digital justice than digital capitalism and capitalist digital platforms. It takes a conscious human effort, social struggle, and material foundations to advance all dimensions of the digital commons. For example, a digital platform can be democratically governed and owned

TABLE 10.4   Three political economies of digital platforms (further development based on: Fuchs 2021, Table 8.2)

Dimension	Capitalist Internet platforms	Public service Internet platforms	Civil society Internet platforms, digital community media
Economy	Digital capital, private ownership of digital platforms that accumulate capital	Public service organisation	Community ownership, civil society organisation ownership, co-operatives
Politics	Governance by private owners, shareholders and managers	Governance by a democratically legitimated board	Governance by the community of members/workers/users
Culture	Publicly available digital content that is prone to ideology and capitalist values	Digital content and digital services that realise the public service remits of democratic communication, education, culture, and participation	Digital content and services that support user-generation, citizen journalism, and digital participation

Conclusion: Advancing Digital Democracy

(political and economic) but advance e-waste and climate change. The organisations and communities operating these platforms should therefore support the creation of non-capitalist green computing.

Political colonisation is the main danger that public service Internet projects face: public service media lose their independence and critical character when governments are able to directly influence the appointment of boards, the hiring and firing of workers, and the produced content. Such media are state-controlled media, not public service media. Just like traditional public service media, public service Internet projects face the danger of political colonisation. Marginalisation and neoliberalisation are the two main dangers that civil society Internet platforms such as platform co-operatives face. The history of alternative and community media is a history of resource precarity and self-exploitative, precarious, voluntary labour. Resource precarity and precarious labour are the two twin political-economic dangers that alternative, community projects face. In addition, digital culture is highly shaped by a culture of individualism and neoliberal entrepreneurialism. The two main dangers that platform co-operatives face are that they (a) remain fair and democratic but small, precarious, and unimportant, which can be their ruin, and (b) that they turn into capitalist projects.

Nick Srnicek (2017, 127) argues that "all the traditional problems of co-ops (e.g. the necessity of self-exploitation under capitalist social relations) are made even worse by the monopolistic nature of platforms, the dominance of network effects, and the vast resources behind these companies" (Srnicek 2017, 127). Marisol Sandoval (2020) analyses how platform co-operatives have employed the neoliberal language of entrepreneurship ("creators", "entrepreneur", "innovation", "investments", "shareholders", "profits", "shares", etc.) and how such a focus has advanced individualism and undermined co-operatives' potential for radical politics. "But collective ownership and democratic governance do not automatically protect co-ops from the dynamics of entrepreneurialism" (Sandoval 2020, 811).

The main danger that public service Internet platforms and platform co-operatives face is that they are paralysed or destroyed by contradictions that stem from economic, political, or ideological colonisation so that they cannot challenge and oppose the power of capitalist Internet platforms. Advancing an alternative Internet can therefore only be successful if it is part of a broader political movement and campaign for strengthening the public sphere and the commons in society. Advancing the digital public sphere, the digital commons, and digital democracy requires progressive politics that address issues such as the following ones:

1) **Techno-realism:** Progressive digital politics should avoid both techno-optimism and techno-pessimism and advance realist projects and platforms that are possible, feasible, challenge, and oppose and point beyond digital capitalism.

2) Advancing digital democracy, the digital public sphere, and the digital commons should be part and parcel of movements, parties, and movement parties that campaign for the **strengthening of democracy, the public sphere, and the commons in general.** Advancing the common control of the means of communication requires the advancement of the common good and the commons in society in general.

3) Advancing the digital democracy, the digital public sphere, and the digital commons is not a technical question but a question of bringing about **good working conditions** of digital and communication workers (and workers in general) and a **good life for all in digital society**.

4) Progressive digital politics needs to stand up for the **breakup of capitalist monopolies** in the communication, media, and digital sector in particular and in the economy in general.

5) Progressive digital politics should demand and advance ending corporate tax havens, corporate tax avoidance, and low corporation taxes. It should campaign for and implement **higher corporate tax rates** in general and in particular a digital services tax that affects large transnational capitalist media and digital companies.

6) Digital democracy, the digital public sphere, and the digital commons need **space, time, material support, and public/civil society partnerships**. Material support helps creating such space and time. Corporation taxes and a media fee paid not just by citizens, but also by companies, can create a material support for alternative projects. The licence fee should be kept where it exists and introduced where it does not yet exist and be used for funding public service media and public service Internet projects. The licence fee should be extended from households to corporations and be developed from a flat fee into a progressive fee. Participatory budgeting can be combined with corporation taxes in order to create a public sphere cheque that citizens receive in order to support alternative, democratic public sphere, and civil society projects. Instead of public/private partnerships, public/civil society partnerships are needed where public organisations co-operate with civil society organisations. Where possible and feasible, there should be partnerships between public service Internet projects and civil society Internet projects. Using such forms of material support, public service Internet

projects and civil society Internet projects, and networks of public service and civil society organisations should create Internet platforms of, for, and through the public sphere that advance the digital commons and follow the remit of advancing democracy, education, culture, and participation in society with the help of digital technologies. Such public, civil, and public/civil Internet platforms challenge capitalist Internet platforms and thereby digital capitalism.

7) **Digital, critical, and democratic skills**: Digital democracy requires critical, engaged citizens who practice democratic debate and democracy. Citizens require time, spaces, educational opportunities, and participation opportunities in order to develop and practice democratic, digital, political, social, cultural, and other skills. On the one hand, participation and engagement with others are education on participation. On the other hand, measures such as the reduction in working hours with full wage compensation, the introduction of a redistributive basic income guarantee funded by capital taxation, political and digital education in schools, and an offensive in adult learning based on the principles of critical pedagogy, etc. are material measures that provide foundations and support for skills development. Progressive digital politics should advance critical education opportunities.

8) **Deceleration, slow media**: The public sphere needs time for critical thinking, reading, critical writing, critical presentation, critical debating, and critical co-production. Digital media can support such processes that blind online practices and face-to-face practices. Digital platforms should be designed in such a way that they enable humans to afford sufficient time for the critical skills just mentioned.

9) **Privacy friendliness and data minimisation**: Non-capitalist digital media should respect the privacy of citizens, workers, and consumers. They should use the principle of privacy by design, minimise data storage to data that is needed for operating platforms, and be advertising free.

10) Public service Internet platforms and civil society Internet platforms as well as their users should **respect and advance democracy,** the plurality of opinions that respects human rights and the equality of all humans, anti-racism, gender equality, anti-fascism, anti-classism, and inclusion. Advancing participation shouldn't be a fig leave for enabling fascist, racist, and other hate speech. News and educational programme require high-quality standards and should always be truthful. There is no place for false news and post-truth politics in progressive media. Those who hold discriminatory views should be allowed to speak as long as they do not violate laws (e.g. when voicing death threat or violent threats), but their views should always be adequately challenged.

Today, digital society is a digital capitalism that undermines democracy, the public sphere, and the common good. Progressive digital politics that advance the digital public sphere and the digital commons along with the public sphere, public services, and the common in general are the active and practical hope for safeguarding and advancing democracy in the age of and in opposition to digital authoritarianism.

Note

1 Global GDP 2019: US$33.426 trillion, data source: https://data.worldbank.org/indicator/NY.GDP.MKTP.CD (accessed on 7 October 2020).

References

Bauwens, Micahel, Vasilis Kostakis, and Alex Pazaitis. 2019. *Peer to Peer: The Commons Manifesto*. London: University of Westminster Press. https://doi.org/10.16997/book33

Benkler, Yochai. 2013. Commons and Growth: The Essential Role of Open Commons in Market Economies. *University of Chicago Law Review* 80: 1499–1555.

Benkler, Yochai. 2006. *The Wealth of Networks: How Social Production Transforms Markets and Freedom*. New Haven, CT: Yale University Press.

Berlinguer, Marco. 2020. New Commons: Towards a Necessary Reappraisal. *Popular Communication* 18 (3): 201–215.

Birkinbine, Benjamin J. 2020. *Incorporating the Digital Commons: Corporate Involvement in Free and Open Source Software*. London: University of Westminster Press. https://doi.org/10.16997/book39

Broumas, Antonis. 2020. *Intellectual Commons and the Law. A Normative Theory for Commons-Based Peer Production*. London: University of Westminster Press. https://doi.org/10.16997/book49

Cody, Sebastian. 2008. After Kelly: After Dark, David Kelly and lessons learned. *Lobster* 55.

Der Standard. 2001. Der "Club 2" ging vor 25 Jahren erstmals auf Sendung. *Der Standard Online*, 5 October. https://www.derstandard.at/story/733146/der-club-2-ging-vor-25-jahren-erstmals-auf-sendung (accessed 7 October 2020).

Euler, Johannes. 2018. Conceptualizing the Commons: Moving beyond the Goods-Based Definition by Introducing the Social Practices of Commoning as Vital Determinant. *Ecological Economics* 143: 10–16.

Fuchs, Christian. 2021. *Social Media: A Critical Introduction*. London: Sage. Third edition.

Fuchs, Christian. 2020a. *Communication and Capitalism: A Critical Theory*. London: University of Westminster Press. https://doi.org/10.16997/book45

Fuchs, Christian. 2020b. *Nationalism on the Internet: Critical Theory and Ideology in the Age of Social Media and Fake News*. New York: Routledge.

Fuchs, Christian. 2018a. *Digital Demagogue: Authoritarian Capitalism in the Age of Trump and Twitter*. London: Pluto.

Fuchs, Christian. 2018b. *The Online Advertising Tax as the Foundation of a Public Service Internet*. London: University of Westminster Press. https://doi.org/10.16997/book23

Fuchs, Christian. 2014a. Digital Prosumption Labour on Social Media in the Context of the Capitalist Regime of Time. *Time & Society* 23 (1): 97–123

Fuchs, Christian. 2014b. Social Media and the Public Sphere. *tripleC: Communication, Capitalism & Critique* 12 (1): 57–101. https://doi.org/10.31269/triplec.v12i1.552

Fuchs, Christian and Marisol Sandoval. 2013. The Diamond Model of Open Access Publishing: Why Policy Makers, Scholars, Universities, Libraries, Labour Unions and the Publishing World Need to Take Non-Commercial, Non-Profit Open Access Serious. *tripleC: Communication, Capitalism & Critique* 11 (2): 428–443. https://doi.org/10.31269/triplec.v11i2.502

Habermas, Jürgen. 1989. *The Structural Transformation of the Public Sphere*. Cambridge, MA: MIT Press.

Hardt, Michael and Antonio Negri. 2017. *Assembly*. Oxford: Oxford University Press.

Harvey, David. 2005. *A Brief History of Neoliberalism*. Oxford: Oxford University Press.

Held, David. 2006. *Models of Democracy*, Cambridge: Polity.

Hess, Charlotte and Elinor Ostrom. 2007. *Understanding Knowledge as Commons: From Theory to Practice*. Cambridge, MA: MIT Press.

Knoche, Manfred 2020. Science Communication and Open Access: The Critique of the Political Economy of Capitalist Academic Publishers as Ideology Critique. *tripleC: Communication, Capitalism & Critique* 18 (2): 508–534. https://doi.org/10.31269/triplec.v18i2.1183

Köhler, Benedikt, Sabria David, and Jörg Blumtritt. 2010. The Slow Media Manifesto. http://en.slow-media.net/manifesto (accessed 2 November 2019)

Marx, Karl. 1894. *Capital. Volume 3*. London: Penguin.

Marx, Karl. 1867. *Capital. Volume 1*. London: Penguin.

Marx, Karl and Friedrich Engels. 1848. Manifesto of the Communist Party. In *Marx & Engels Collected Works (MECW) Volume 6*, 477–519. London: Lawrence & Wishart.

Papadimitropoulos, Vangelis. 2020. *The Commons: Economic Alternatives in the Digital Age*. London: University of Westminster Press. https://doi.org/10.16997/book46

Rauch, Jennifer. 2018. *Slow Media. Why "Slow" is Satisfying, Sustainable, and Smart*. Oxford: Oxford University Press.

Rosa, Hartmut. 2013. *Social Acceleration. A New Theory of Modernity*. New York: Columbia University Press.

Sandoval, Marisol. 2020. Entrepreneurial Activism? Platform Co-Operativism between Subversion and Co-Optation. *Critical Sociology* 46 (6): 801–817. https://doi.org/10.1177/0896920519870577

Scholz, Trebor. 2017. *Uberworked and Underpaid: How Workers Are Disrupting the Digital Economy*. Cambridge: Polity Press.

Scholz, Trebor. 2016. *Platform Cooperativism: Challenging the Corporate Sharing Economy.* New York: Rosa Luxemburg Stiftung New York Office.

Scholz, Trebor and Nathan Schneider, eds. 2016. *Ours to Hack and to Own: The Rise of Platform Cooperativism, a New Vision for the Future of Work and a Fairer Internet.* New York: OR Books.

Splichal, Slavko. 2007. Does History Matter? Grasping the Idea of Public Service at its Roots. In *RIPE@2007: From Public Service Broadcasting to Public Service Media*, ed. Gregory F. Lowe and Jo Badoel, 237–256. Gothenburg: Nordicom.

Srnicek, Nick. 2017. *Platform Capitalism.* Cambridge: Polity Press.

Utman, Jorge S. 2020. Subversive Communication against Neoliberalism. *Popular Communication* 18 (3): 155–169.

Žižek, Slavoj. 2010. How to Begin from the Beginning. In *The Idea of Communism*, ed. Costas Douzinas and Slavoj Žižek, 209–226. London: Verso.

References

Index

Note: **Bold** page numbers refer to tables; *italic* page numbers refer to figures and page numbers followed by "n" denote endnotes.

For Product Safety Concerns and Information please contact our EU
representative GPSR@taylorandfrancis.com
Taylor & Francis Verlag GmbH, Kaufingerstraße 24, 80331 München, Germany

www.ingramcontent.com/pod-product-compliance
Lightning Source LLC
Chambersburg PA
CBHW050347270326
41926CB00016B/3631

9 781032 246161